Veniamin Mikhailovich Tarnovskii

The Sexual Instinct and Its Morbid Manifestations

From the Double Standpoint of Jurisprudence and Psychiatry

Veniamin Mikhailovich Tarnovskii

The Sexual Instinct and Its Morbid Manifestations
From the Double Standpoint of Jurisprudence and Psychiatry

ISBN/EAN: 9783337312244

Printed in Europe, USA, Canada, Australia, Japan

Cover: Foto ©berggeist007 / pixelio.de

More available books at **www.hansebooks.com**

THE SEXUAL INSTINCT

AND ITS

MORBID MANIFESTATIONS

EDITION

STRICTLY LIMITED

TO SEVEN HUNDRED AND

FIFTY COPIES WITH

PORTRAIT OF THE

AUTHOR

S. Tarnousky 16/25 III - 98

Professor Benjamin Tarnowsky

THE SEXUAL INSTINCT

AND ITS

MORBID MANIFESTATIONS

FROM THE DOUBLE STANDPOINT

of JURISPRUDENCE and PSYCHIATRY

BY

Dr. B. TARNOWSKY

(IMPERIAL ACADEMY OF MEDICINE, ST. PETERSBURG)

TRANSLATED BY

W. C. COSTELLO, PH.D. and ALFRED ALLINSON, M.A.

PARIS
CHARLES CARRINGTON
PUBLISHER OF MEDICAL, FOLKLORE AND HISTORICAL WORKS
13, Faubourg Montmartre
MDCCCXCVIII

PREFACE TO THE ENGLISH EDITION

Medicine undertakes to save the honor of mankind before the Court of Morality, and individuals from judges and their fellow-men. The duty and right of medical science in these studies belong to it by reason of the high aim of all human inquiry after truth.

(Dr.) KRAFFT-EBING.

PREFACE TO THE ENGLISH EDITION.

In the present work I sketch in broad outline those facts and observations which inspired me with the idea of making enquiry into the causes of sexual perversion,—and this not merely under the influence of depravity and licentious excess; preferably, in fact I may say chiefly, I examine these causes as connected with a morbid condition of the organism, whether congenital or acquired.

Above all I make it my business to throw all the light possible on the part played by heredity and by the phænomena of arrested development, as well as by the various morbid causes conditioning the etiology of sexual perversions, and to differentiate these with the very utmost clearness from proved depravity of character—conscious and premeditated vice.

My Treatise on Sexual Perversion appeared in the first instance in Russian in 1885. A large number of Works have followed suit since in different countries, and I have enjoyed the very great satisfaction of noting that my conclusions have been in the main confirmed by all my learned fellow-workers in other parts of Europe.

Carrying my investigations further into this subject, one equally delicate and important, I have since brought together a very considerable number of fresh observations, all of which support the views originally expressed in the present work. But, as previously to last year the whole of my time was consecrated to teaching my Classes at the Academy of Medicine, leisure has hitherto failed me to draw up a fair and proper statement of my observations. From another point of view, it may be that, considering their special subject, they are still too recent to be published at once.

Meantime I am bound to state that this fresh evidence has not in any way modified my convictions on the question of sexual aberrations. I may add that the further experience of these last few years makes me insist with even greater confidence than before on the practical conclusions with regard to examination by medico-legal experts which I had previously drawn, and which are laid down in this book.

I would beg my excellent translators, Messrs. W. C. COSTELLO, Ph.D., and A. R. ALLINSON, M.A., as also Mr. CHARLES CARRINGTON, my Publisher, to accept my very sincere thanks for their kindness in undertaking the English Translation. I owe it to them that I am in a position to bring my book under the notice of my English colleagues, as well as that of English Jurists,— Physicians and Jurists being the two classes I had particularly in view when I wrote my book.

Professor BENJAMIN
TARNOWSKY.

ST. PETERSBURG,

  7 March, 1898.

CONTENTS

Now that the Problem of religion has practically been settled, and that the Problem of labour has at least been placed on a practical foundation, the Question of sex—with the Racial questions that rest on it—stands before the coming generations as the chief problem for solution. Sex lies at the root of life, and we can never learn to reverence life until we know how to understand sex.

HAVELOCK ELLIS.

> "Einstweilen bis den Bau der Welt
> Philosophie zusammenhält,
> Erhält sie das Getriebe
> Durch Hunger und durch Liebe."
>
> SCHILLER.

CONTENTS.

PREFACE TO THE ENGLISH EDITION VII
INTRODUCTION XVII
GENERAL CONSIDERATION 1

Group A. PERVERSION OF THE GENESIC INSTINCT BASED ON HEREDITARY INFIRMITY.
   I. CONGENITAL CONTRARY-SEXUAL FEELING (CONGENITAL PEDERASTY) 8
   II. PERIODICAL PERVERSION OF THE GENESIC INSTINCT (PERIODICAL PEDERASTY) 52
   III. SEXUAL PERVERSION OF EPILEPTICS (EPILEPTIC PEDERASTY) 71

Group B. ACQUIRED GENESIC PERVERSION.
   I. ACQUIRED PEDERASTY 91
   II. GENESIC PERVERSION IN THE DEMENTIA OF DOTAGE (SENILE PEDERASTY) 106
   III. GENESIC PERVERSION IN PARALYTIC DEMENTIA (PEDERASTY IN THE PROGRESSIVE PARALYSIS OF THE INSANE) 118

Group C. COMPLEX FORMS OF GENESIC PERVERSION 129
BIBLIOGRAPHY 225
INDEX 233

INTRODUCTION

Exact knowledge of the causes and conditions of development of sexual aberrations, and of the influence on them of hereditary constitution, education, the impressions of every-day life, and modern refined civilization, is the prerequisite for a rational prophylaxis of sexual aberrations, and for a correct sexual education.

(Dr.) SCHRENCK NOTZING.

INTRODUCTION.

Five years ago I was called upon to give my opinion as an expert in a case of pederasty.

On looking through my observations on this subject, and comparing them with the corresponding chapters of the most widely known manuals of forensic medicine, I was struck with the want of agreement between the assertion of official science on the one hand and clinical facts on the other. Each fresh technical examination taught me more clearly to recognize not only how insufficient was the knowledge contained in the manuals dealing with perversity of genesic activity, but also the incorrectness of many of the guiding principles of examination into the actual circumstance. Later studies on this matter by KRAFFT-EBING, LOMBROSO, CHARCOT, MAGNAN and other alienists have rendered it quite impossible to adhere to former views, and yet further confirm my conviction of the truth of the conclusions I have deduced from clinical observation.

At present the difference of opinion between the medical jurist and the clinical physician has become so broad, that it appears to be highly necessary, in order to explain these contradictions, to indicate the foundations on which approximately correct answers to medico-forensic questions may be based, and to prepare the way for wider, more united and more fruitful study of the subject.

The medical jurist sees depravity, over-satiated lust, inveterate vice, wickedness, and so on, where the clinical observer recognizes with certainty the symptoms of a morbid condition with its typical evolution and result. Where the former would punish vice, the latter enters a plea for the necessity of methodical therapeutics. On the other hand, there is a whole series of acts, which are relatively but lightly punished by the law, in which the medical jurist sees only impropriety, caprice or play carried too far, recognizing all the time the moral responsibility of the culprit, whereas the clinical observer discerns therein the outset of a grave malady, the beginning of an incurable psychical disturbance, requiring careful watching and treatment. Lastly, the clinical doctor discovers real depravity and complete moral decay in cases, where the medical jurist is more often inclined to suppose the wrong-doer a victim of violence or of fraud.

It is certainly not difficult to understand, why, in any question as to perversity of the genesic activity, the observations and conclusions of the clinical physician must take precedence in the discovery of the truth. The medical jurist has exclusively to do with incriminated persons, who first of all seek to escape punishment and therefore mostly deny quite obvious facts. Even in the rare cases of frank admission of culpability the accused sees no reason for describing the intimate causes and motives of his actions. The physician appears naturally in his eyes more as an accuser than as a defender; and when, having made up his mind to undergo the sentence, he confesses the facts, he has no object to gain in making an avowal, one generally fraught with ill-consequences to himself, of his perverted sexual instincts and the impulses that drive him to gratify his genesic sense in abnormal ways. It is but seldom that the medical jurist can obtain the diaries of such individuals, their correspondence or other interesting documents, such as are for instance to be found in the classical treatises of Caspar and of Tardieu. But such diaries and autobiographies show very prominently the usual fault of all productions of the kind, viz. the wish to excite interest, to exhibit their good aspects only. In this way the description becomes exaggerated, untrue and utterly deceptive. The clinical physician on the contrary has not to examine incriminated persons. The sufferer comes to him spontaneously to seek for advice. He hopes for assistance from the physician and in all sincerity confides his ailments to him, even in their slightest details. There is no room here for premeditated fraud or conscious unveracity. The physician is therefore in a position to distinguish the initial symptoms of an anomaly or of a malady, to follow its development and observe its progress during a number of years. By such means he obtains a complete, consistent and decisive view of the disease.

The representatives of two specialities are the most frequently consulted by such patients:

physicians devoted to the treatment of mental diseases and specialists for diseases of the genital organs. To the first the applicants are mostly persons with well-defined symptoms, in whom there also exists concurrent perversity of the genesic instinct, not to mention other derangements of the nervous system. The disturbance of the genesic activity is here only an accessory symptom among the series of the other more serious cerebral phenomena which trouble the patient. The specialists for mental diseases have besides opportunities for observing in their asylums the ultimate form of those maladies, which at their inception are manifested in perversity of the genesic action and terminate in complete madness.

Specialists for diseases of the genital organs are resorted to by such as find themselves suffering from any manifestations of organic anomaly, however slight, from arrested development or from the premonitory signs of incipient disease,—when a diminution of the sexual power comes into prominence above all other symptoms, or finally when syphilitic infection makes an avowal inevitable of some perversity of the genesic sense of one sort or another.

Amongst those consulting the specialist of this class are: The Young Man who after attaining the age of manhood becomes aware of complete impotency where women are concerned, and is obscurely conscious of abnormal sexual promptings within; the Old Man, whose sexual activity has long ago disappeared, but who suddenly feels a new sensual impulse, an exaggerated wantonness and morbid stimulation of desire; the Husband, who idolizes his wife and from time to time gives way to the irrepressible sexual impulse, but who feels impelled to accomplish the conjugal act in some quite unusual and to his own consciousness disgusting manner; the Voluptuary, who has become aware of a diminution of his genesic power and does not know whether it is to be attributed to a quantitative or a qualitative alteration of function. All these haste first of all to the specialist in maladies of the genital organs for advice. Again it is to him the habitual pederast makes his involuntary confession, who has accidentally become inoculated with syphilis, and the boy who has recently been seduced and finds himself attacked by some trouble of the anus or rectum. The avowal of their failing is usually a source of great moral mortification to patients. It is often made in writing, with precautions of mystery and secrecy; and then a frank confession can only be expected, if the physician avoids meeting it with reproof, but holds himself ready to give his assistance.

I may further remark in this place, that nearly all pederasts, of whatever rank, are more or less known to each other, at least in large towns, and that they generally go for advice to the same physician, whose task of obtaining open and frank confessions with regard to such abnormal genesic facts is much facilitated by the fact.

In the course of my 25 years of medical practice, I have had more particularly to do with various phases of disease affecting the genital organs. And as I carefully noted all my observations I was able to make a large collection of facts bearing upon morbid manifestations of the genesic sense.

The actual criminal and the undoubted madman are two extremes, between which is found a host of unrecognized sufferers, and vicious subjects burdened with an abnormal function of the genesic action; of these two classes the latter will furnish the greater part of the material for the following investigation.

I trust that observations, not emanating from the prosecuting bench, nor from the registers of lunatic asylums, observations taken on persons belonging to society in general, who have not been bereft of their legal status, and who may be held to be sane, may supply new data for a proper differentiation between vice and disease, between congenital defect and moral lapse.

My intention has been to make my treatment of the subject intelligible not only to the physician, but also to the jurist. Therefore the exposition may be somewhat lengthened, because of the necessity of digressions in order to make it as generally comprehensible as possible. I am convinced that it can only be by an exact enquiry into so-called offences against public morals, and a methodical exposition of the facts as compared with the usual legal procedure, that will enable jurists, thoroughly initiated into the actual state of science relating to the morbid manifestations of the genesic sense, to come to a proper conclusion.

GENERAL CONSIDERATIONS.

The genesic instinct presents itself in the human being previous to the development of sexual maturity, and in a healthy organism finds expression in an impulse towards persons of the opposite sex, an impulse which soon changes into a voluntary desire for the accomplishment of the sexual act which, as it becomes more developed, may attain to the degree of a continual sexual want.

Among anomalies of the genesic functions perversity of the genesic instinct shows itself most distinctly in a morbid inclination towards persons of the same sex. Among men, who form more exclusively the subject of the present research, this form of sexual perversity is designated by the comprehensive expression of "Pederasty." [1] Bestiality and Sodomy are,

---

[1] Formed of two Greek words, παιδὸς ἐραστῆς, i.e. *pueri amator;* better known to the Romans under the denomination of "Grecian Love."

as we shall see further on, no more than variations of the above-mentioned abnormal genesic impulse.

The pederastic manifestations constitute the more definite and better studied group of sexual aberrations, as they are more often the subject of judicial investigation; and for this reason the most prominent place is given in the present work to Pederasty in the widest sense of the expression.

But this vice must not be isolated in this study, without considering also its connection with the other forms and kinds of perversion of sexual activity, which we shall therefore also touch upon, as much as may be necessary for the elucidation of the subject.

As Pederasty is developed under the influence of very different causes, it must, as all other aberrations of the genesic sense, be distributed from the clinical and etiological points of view into several groups and kinds, differing considerably one from another.

The morbid manifestations of sexual activity separate first of all into two great groups, according as they develop in such subjects as are from their birth disposed to such aberrations and to nervous disorders of all sorts, or on the other hand, show themselves in persons, relatively healthy, and who are not under the influence of hereditary infirmity.

In the *first group* the disorder of the genesic functions is to be referred to a psychopathic or neuropathic constitution; whereas in the second group we have to do with individuals, who from their birth have always enjoyed a properly constituted and normally acting nervous system.

The morbid symptoms again resulting from hereditary infirmity also show themselves under different forms.

In some cases they present themselves with the very first awakening of the genesic instinct, and in their development remain refractory to the influence of education or example, attain their greatest intensity in the period of sexual maturity and manhood, and subsist during the rest of the lifetime, with periodical diminution or increase.

These aberrations are in their nature constant and invariable, special to the particular organism, in precisely the same way as other innate peculiarities of character feeling or morality are special to the same individual.

The said disturbances constitute the first kind of hereditary infirmity, which we will denominate in general *innate Perversion of the sexual instinct;* and in this group we shall class *innate Pederasty.*

To the second division of the same group belong such modifications of the genesic activity as now and then show themselves *nolens volens* in the form of morbid attacks, which are separated by intervals, during which the sexual functions are performed in a normal manner.

The denomination *periodical Perversion of the genesic sense* in general, and *periodical Pederasty* in particular, appears to me best to represent the morbid disturbances in question, because they are in perfect analogy with the so-called periodical psychoses (mental disturbances), those higher manifestations of psychical degeneracy,—a degeneracy manifesting itself in occasional outbursts of mental disturbance, while there are long intervals between the attacks during which the mental powers exhibit a relatively normal activity.

In a third class of genesic disturbances I reckon those disorders which show themselves in the well-known morbid manifestations known under the general name of Epilepsy.

During the mental disturbance incidental to this malady, there may be exhibited a sort of psychical epilepsy, in the form of an epileptic pederasty.

I regard as paroxysmal phenomena analogous to hysteria and mania those morbid manifestations of the genesic sense that are observed in the so-called erotomania and satyriasis. The description of the last named forms the termination of the group of sexual perturbations founded on hereditary infirmity.

The *second group* includes all those aberrations of the genesic activity, that may be the result of education or example, or come spontaneously from personal impulse, as the expression of vicious propensity, or of wilful depravity.

The expression: "*Acquired Perversion*" *of the sexual functions and acquired Pederasty* answers best in our opinion to the class included in the above group.

This group also contains sundry subdivisions consisting of such perturbations of the sexual functions as are the symptom of a developing disease of the nervous centres, or of the entire organism—a disease that attacks subjects who from their birth have had a healthy, well constituted brain.

Among these are to be reckoned the sexual perversions peculiar to the decrepitude of old age—*Senile Pederasty;* and again the sexual aberrations observed during the initial period of paralytic idiocy—*Paralytic Pederasty.*

Every student of nature knows, that the division into groups and species based on any particular symptom is purely conventional and artificial, and serves only to facilitate research without being grounded on any immovable, unchangeable organic foundation. The same holds good with the grouping proposed by me. It must needs undergo modifications as life goes on; the various kinds and types little by little dissolve into each other, becoming mutually complicated one with the other, acquire new, fresh shades of colouring and form, according to the composite forms of genesic aberration, which will be described as a separate group.

I have further analysed the objective symptoms, which enable one to form an opinion on the pederastic perversion of the sexual activity, and which must needs constitute the basis of any special medico-judicial investigation.

Finally, I have endeavoured to determine the data which may serve to establish a differential diagnosis of the different groups and classes of sexual aberrations.

We shall begin by describing the forms of sexual aberrations most frequently encountered.

GROUP A. PERVERSION OF THE GENESIC INSTINCT BASED ON HEREDITARY INFIRMITY.

I. Congenital Contrary-Sexual Feeling (Congenital Pederasty).

JUST as children may be born with abnormally constituted extremities, trunk, head or other members, a congenital tendency may in like manner appear towards perverse modes of manifestation of the genesic instinct.

As in a physical cripple the whole organism may be more or less normally developed, apart from the particular abnormality, so in a moral abortion the psychical constitution may be more or less normal, with the exception of the particular perversion present. The irregular or faulty development of the nervous centres may present a particular character, by reason of which a functional disturbance caused thereby may be limited to any particular domain, that for instance of sexual activity.

But the faulty development of certain nervous centres is not without influence on the development of the others. It is generally possible to recognize in an abortion, notwithstanding the more or less regular constitution of the non-affected parts of the body, a number of slight deviations and irregularities which, if not apparent on a superficial view, are nevertheless discovered by careful examination. In the same way where there is congenital disturbance in limited nervous centres, it has a certain influence upon the whole of the nervous system, which shows itself in certain hardly perceptible irregularities. In order therefore to obtain a general view of the anomaly in question with all the irregularities connected with it, it is necessary to follow up its general development from its very first manifestations.

The child born with congenital sexual perversion grows up and develops to all appearance quite regularly in every way. The genesic sense alone generally awakens unusually early and, as the period of sexual maturity approaches, a whole series of abnormal morbid manifestations show themselves. The first of these symptoms is an expression of shame, not before girls or women, but in presence of grown-up men. For instance, the boy is more ashamed to undress himself before a strange man, than before a woman. On the other hand he prefers the company and caresses of men to those of women. He feels a great attachment to a man, follows him about incessantly, obeys him without a murmur, is charmed with him—in one word, he "adores" a man, either a brave, generous and clever man or one with strongly developed muscles, whilst he remains quite indifferent towards women. At length he attains puberty; in the night he often has violent erections with emission of semen. The pollutions are accompanied by dreams, at first indetermined, easily forgotten; but they each time become more distinct, decided and often astonish the youth himself by their strange nature. They are not female caresses nor meetings with women that appear to him in his dreams, but the pressure of the hand, the kiss of well-built, handsome grown-up men. The final erection ending with seminal emission is not provoked in dream by a female form in seductive attitudes and movements, but by the embracing, caresses and kisses of men.

Not only does the representation of womanly form give rise to no sexual excitation, but it paralyzes all voluptuous feeling, when such exists already. Among normally constituted men the heat of sexual erethism rapidly disappears at the sight of grown-up men. On the contrary this same erethism is quenched at once in the congenital pederast in the presence of women. The sight of a naked young woman leaves him indifferent, whereas that of a naked man will awaken in him feelings of lust.

From books and conversation with his comrades he discovers that something quite unusual and abnormal is taking place within him. But the strangeness to him of the phenomenon itself, the difficulty of giving it a distinct form, and his ever increasing shyness when in company with men, cause the youth to dissimulate his misfortune. Sometimes, incited thereto by companions of his age, he ventures to share the couch of some girl and to accomplish the act of manhood, but each time the effort fails and is not seldom followed by a hysterical fit.

Just as a normally constituted and sexually developed man, try as he may, cannot feel any lustful desire for another man, so it is equally impossible for the congental pederast to accomplish coition with a woman. Such fruitless endeavours serve only to still more discourage the young man and finally inspire him with a disgust for women. He now seeks the companionship of men, pays court to them, falls in love with them, and in the meantime seeks for satisfaction in onanism.

In consequence of the congenital morbid erethism, the so-called excitable weakness of the nervous system, the feelings of lubricity become soon intensified to their utmost, and often find in the simple physical contact with the beloved person the determination of an *emissio seminis*. The love of such subjects is extraordinarily violent, morbidly passionate, absorbs all their moral and sensitive faculties, and is at first purely Platonic; it is only later on that the sexual feeling and its consequent genesic impulse finds satisfaction in mutual masturbation or in onanistic excitation of the adored person. Finally, when the latter consents to it, or because he wishes to satisfy his lust otherwise, the act of sodomy is consummated, in which the morbidly disposed participator always plays the part of the passive pederast.

Concurrently with the exaggeration of sexual perversion just described, other peculiarities of the sick organism begin to show themselves. The youth is impelled to give himself a feminine appearance, likes to wear female attire, to have his hair curled, to walk about with exposed neck, and tightened waist, to scent, powder and rouge the face, to paint the eyebrows, etc. In this way is developed a type of womanish-looking men, disgusting to those of their own sex and looked upon with contempt by women, and whom it is easy to recognize by their appearance. They are generally of middle-sized or slight build, with broad hips and narrow shoulders, affecting a feminine walk and a peculiar swinging gait; with perfumed locks, eccentric dress, bracelets on their wrists, they seek by every means, laughing, talking and gesticulation, to draw the attention of men to themselves. The unhappy creature cannot realize, particularly if there is a relatively weak development of his understanding, that he is so much the more repugnant to normally constituted men the more he seeks to imitate a woman. At once fantastical, even to hysteria, envious, cowardly, mean, revengeful and spiteful, he combines in himself all the faults of the woman, without any of her qualities, and possesses none of the attractive traits of the manly character. He is therefore despicable alike to men and women.

Many such sufferers willingly acknowledge their abnormal condition which they endeavour to explain to themselves. "I have a female soul in a man's body," wrote K. H. Ulrichs, [1] who has written quite a series of treatises, which, although very interesting as the detailed confessions of a psychopathic subject, are at the same without order and too long drawn out, as is the case with most of the same class of works.

It often happens that such patients feel depressed in mind by the knowledge of their infirmity and of their inability to fight against it. In this connection, a letter, kindly communicated to us by Professor *J. Mierzejewsky*, is of great interest. It is to the following effect: " .... To speak frankly, I am even here, far away, exposed to temptations, against which I am defenceless, and I do not really know what it means; malady, or the power of youthful impressions, or want of will, against which I have been fighting in vain for more than four years—or an unlucky

---

[1] *Numa Numantius*, Anthropologische Studien über die mannmännoliche Geschlechtsliebe, etc. (Anthropological Studies of Sexual Love as between Man and Man), Leipzig, 1869.

nature, a fatal destiny, or else an instinct
that death only can cure?! That is why I
at times brood on the thought of some rapid
and painless poison, for in truth this
instinct is in contradiction with my
judgment...."

A few months before his marriage with a
charming girl, a patient of mine wrote to
me as follows: "I am a slave to my fatal
passion, I cannot abandon vice and I am not
able to love Fräulein X..., although I feel
that it is only with her that I could be
happy. In her absence, I love her understanding, her soul, her visage even; but when
I see her, I feel that I should not be in
a condition to become her husband.... It
would be the greatest of misfortunes. Up
to the present I have never been able to
have intercourse with women—with her it
will be the same thing... There remains
for me nothing but death, if your help proves
to be powerless..."

Sometimes the patient, perpetually tortured
by jealousy of women because of their success
with men, feels despair on account of his
mostly unsuccessful loves, often repulsed with
contempt, in a fit of melancholy takes his
own life or, sinking deeper and deeper, he

confines himself to the narrow circle of a few fellow-sufferers, and terminates his existence in a half stupid condition. Otherwise he may become the victim of one form or another of acute mania, if his miserable existence is not previously put an end to by some other accidental malady. The type here described combines in it the principal characteristics of the congenital pederast as most frequently met with.

However, there are others again who, in their efforts to ape the fair sex, adopt more eccentric forms. So, for instance, Taylor [1] in this connection has written a most important observation, embodying his description of the celebrated English actress and adventuress Elisa Edwards who, after her death, was discovered to have been a man in disguise. From his earliest youth he felt an inclination for men, from the age of 14 wore female clothing, went on the stage, had many amorous adventures, lived as a Mistress, and used to fasten up his genitals, which were quite normally developed, to his body by a special bandage, so as not to be recognized.

Notwithstanding the clear, explicit expression of the inclination to copy the female

[1] Medic. Jurisprudence, 1873. Vol. II, pp. 286 and 473.

sex in the above case, the sexual perversity implied differs really in no respect from that exhibited by the previously described type of the born passive pederast, or so-called "Cynede".

But besides such cases as this, there exist other intermediate forms occurring under circumstances to be presently defined,—morbid aberrations now more, now less, strongly marked, which form a gradual transition to obvious forms of hereditary mental derangement.

In the weaker defined manifestations the boy or youth exhibits only his predisposition to occupy himself with feminine work—He likes to knit, to sew, to make doll's clothes; otherwise he distinguishes himself by his peculiar preference for feminine manners, he strives to be graceful and coquettish in his demeanour, imitates the tone and voice of a woman, etc., and is awkward and out of his element and blushes when in conversation with men. On the contrary, he is quite free from embarrassment in the company of young girls, is glad to take the woman's part in dancing, always choosing for cavalier a vigorous, stalwart manly dancer, becoming quite enlivened and merry, when he has found a man to his taste, or else growing confused and troubled at his sight and running away like a timid little girl.

Another occupies all his leisure time before the looking-glass; combs his hair, puts on curl-papers, paints his face, adorns his person, studying in the most serious fashion what is becoming to him and what is not. He has a wonderful remembrance of the most complicated female toilettes, and is able to describe them in all their details; in this matter he exhibits a quite delicate taste, but shows himself absolutely wanting in taste when he adopts male attire. He either sports a too violently coloured neck-tie, or he exposes his neck so low, that it appears extravagantly exaggerated, even for a woman; or else he has his hair curled in long locks, and covers his fingers with rings and puts bracelets upon his wrists.

In other less decided forms, the sexual instinct is more frequently and particularly determined by the surrounding circumstances. When the parents or the instructors of such morbid subjects fail to comprehend the signification of the above-described manifestations, give them only a frivolous consideration, and even for fun encourage such feminine instincts, the inevitable result is that the morbidly disposed youth generally becomes an onanist, not feeling attracted towards the female sex, from which he more and more estranges himself, and finally on favourable occasion becomes a pederast, although, at first, he still possessed the power of sexual intercourse with women.

The more intense the morbid manifestation, the longer has the subject been addicted to masturbation, and the sooner has he become a pederast, and the sooner also does he lose the possibility of normal coition.

Yet more fatal in its effects upon subjects that way disposed is the companionship of similar morbidly affected comrades, in a still more advanced pathological condition, as often occurs in schools. By the example of elder comrades the boy early becomes a pederast and consequently at the advent of puberty his morbidly diminished desire for the female sex is still more accentuated.

Under more favourable circumstances the case has a more satisfactory culmination.

When the boy has been repressed in time, and laughed at on the first feminine imitations, he involuntarily begins to pull himself together. If he is then carefully kept away from female society, occupied as much as possible with athletic exercises, always severely reproved and punished for the slightest appearance of coquetry, graceful manner, extravagant delicacy and in general for every external feminine manifestation, by such strictly conducted education the youth attains to the normal state of puberty.

The morbidly diminished sexual inclination towards the female sex that is congenitally present, and the weakened and perverted genesic instinct—the consequence of bad surroundings and education—cause the youth in this initial period of his life to be more indifferent to sexual enjoyment than are his comrades of the same age. It frequently happens that, when he has come to manhood, after violent erethism or repeated pollutions, his first attempts at sexual intercourse with women are abortive; or that, notwithstanding their successful accomplishment, he has not found therein the same enjoyment a normally constituted being would.

However, if he perseveres in having regular intercourse, particularly with one and the same person, the genesic perversion gradually dies out, and finally the youth who from his birth was disposed to perversion of sexual instinct, grows up to be a man endowed with normal genital functions, and fit to fulfil the duties of the head of a family.

Another variety of these morbid manifestations consists in those cases, in which the touching of the hind-quarters determines a sexual erethism, the gratification of which is not perverse and can happen normally. Sometimes the boy notices in quite early youth, that slight strokes on his naked posteriors caused him an agreeable sensation.[1] He then voluntarily seeks, in play, in joke, or even as a punishment to get a few strokes. When this particular predisposition is not taken notice of, the strokes, especially those of the birch, awaken erotic excitement in the boy. Later on he fustigates himself, when he is alone, and the erethism culminates in onanism. When the period of manhood arrives, if the vicious habit, of seeking excitation in flagellation, that is to say by strokes with a birch on the posteriors, has become deep-rooted, the patient is only then able to have intercourse with women, after having been flogged previously, which for ever deprives him of the possibility of family life, and necessarily reduces him to the extremity of

[1] J. J. Rousseau, Les Confessions, Partie I, Livre I.

having recourse to masturbation or to the exclusive frequentation of prostitutes, women who sell their favours. Under such conditions the vicious propensity is still further developed. Strokes alone, even when followed by blood, are not sufficient for the patient; he requires a certain amount of violence to be exercised upon him. He must be undressed brutally, or his wrists are to be tied together, he must be fastened down to a bench, etc., during which he makes a pretence of resistance, shouts and swears. It is only by such means and flogging with birch rods that he succeeds in obtaining that degree of sexual excitation which ends in *emissio seminis*. In this morbid period he seldom goes as far as actual coition, more frequently the semen is expended even without erection—At last the patient loses altogether the faculty of performing the genesic function in a normal manner, and gradually is developed in him the predisposition to graver forms of nervous and mental disease.

It goes without saying that, when the morbid propensity is discovered early, and abnormal excitement by touching of the posteriors is carefully avoided, the period of puberty is thereby retarded as much as possible, and the efforts to overcome the congenital infirmity become more easy and more successful.

Besides the weak congenital forms, which indicate a more or less slight disposition to pederasty, there are other more violent forms, happily less common, which show a gradual transition to complete madness. Many of these subjects find their first feelings of lust excited by the sight of a naked man, particularly of his posteriors or the *orificium ani*. Dr. Albert mentions for instance cases, in which certain schoolmasters whipped their pupils without any cause, the sight of the naked buttocks of the children producing in them a state of sensual excitement which they then satisfied by masturbation.[1]

When such subjects give themselves up early to masturbation, they seek to excite themselves by touching or rubbing against the posteriors of men, on which occasions there is often an emission of semen. They have also nocturnal pollutions accompanied by dream-pictures in which naked men with strongly developed hind-quarters play the principal part. Such subjects are born active pederasts; they are always indifferent to women, have no feminine propensities, but

[1] Albert, Friedreich's Blätter, 1859, III, p. 77.

generally exhibit, besides their sexual perversion, other morbid symptoms as well, indicating a greater or less degree of degeneracy. Some display from infancy an inclination to objectless thievery; others are subject to epileptic fits, with temporary loss of consciousness; others again are afflicted with intellectual dulness, are soon fatigued by any mental labour, are slow to understand things, have a poor memory, and so forth.

On proceeding one step further in psychical degeneracy, we come to subjects who have an exclusive taste for old men. Many born pederasts feel attracted only by men with gray beards. For them neither youth, nor elegance of bodily form, nor beauty, whether in woman or in man, is of any importance; their sexual instinct can only be excited by the aspect of a gray beard, sometimes indeed by the ugliest face, rendered repulsive by deformity.

Another degree of this congenital perversion is manifested by individuals in whom the sexual excitement is produced by the aspect of lifeless objects which have no connection whatever with the sexual act. Cases are known, in which the sight of a nightcap on the head of a man, or of a woman, has caused this erethism with emission of semen. The view of a naked woman or man left the subject indifferent, but the remembrance of a nightcap, particularly if on the head of an old shrivelled-up woman, or the touch of a nightcap, caused immediate erection and even emission.

Another unhappy patient had from early youth been in the habit of directing his attention to the nails in the shoes of women. The contemplation of such shoe-nails procured him particular pleasure; during the night he would get up, secretly seek out such shoes, count the nails and, lying afterwards in his bed, would give himself up to all sorts of phantastic ideas, in which the benailed shoes led him to erection and precocious onanism. Later on the mere sight of a cobbler hammering nails into the soles of the patient's own boots was sufficient to cause him to emit semen without erection.

In a third case the first sexual erethism was caused by the sight of a white apron hung out to dry in the sun. He took the apron down, hung it before him, and commenced masturbating. From that moment the sight of a white apron always caused him erethism. It was, however, indifferent whether the apron was worn by a man or by a woman, or whether it hung on a clothes-line, the sight of the apron invariably awakened in him the irrepressible wish to seize it and masturbate himself with it.

After he had been several times punished for stealing aprons, he entered a monastery, where he sought of his own accord to conquer his flesh by fasting and prayer, but he was not able to become master of his passion, and the remembrance of white aprons was sufficient to make him fall back into his vice. [1]

Another experienced violent lust, terminating in spasmodic erection, when his genitals came into contact with fur, which occurred by accident, when he happened one night to cover himself with a fur rug. From that time—from the age of twelve—he was given to masturbation. The sight of the body of a man or of that of a woman, even during copulation, would not excite him in the least, but to touch a hairy little dog, which he sometimes took into his bed, always caused him erection, ending in emission: this was sometimes followed by a hysterical attack, accompanied by convulsions, sobbing, etc.

[1] Charcot et Magnan, Inversion du sens génital. Arch. de Neurologie, 1882.

His nocturnal pollutions were accompanied by dreams, in which neither men nor women played a part; he used to dream, that he was stretched naked upon a soft fur, that every point of his body pressed amorously to it, and this sensation led to erection and pollution. As he grew older, and became aware of his morbid condition, he was at times in actual despair and was often on the point of committing suicide.

He was very soon fatigued by mental work, and failed in his examinations at the University; his memory became much impaired; it always seemed to him as if his comrades regarded him in a peculiar manner and despised him. It was this last that troubled him most. He used to enquire, if it were possible to recognize his condition by the expression of his eyes? if there existed no means, by the absorption of remedies, to dissimulate his situation from "*them*"? He would cry, and sob, suffered terribly, praying to be saved from himself. "Should I find out, that 'they' guess what is going on within me, I would most certainly kill myself."—These were his parting words at our last interview. Evidently the mania of being persecuted was already developing itself in him. As I was subsequently informed, he was a few months later placed in a lunatic asylum.

This patient exhibited degeneracy to a high degree. His genitals were irregularly developed, and in the bony structure of his body there were distinctly marked malformations. But similarly, in all previously cited cases, there had always been observed, besides a high degree of perversion of the sexual instinct, other manifestations as well, which indicated an abnormal development of the nervous system caused by degeneracy. Symptoms were always present of a morbid, or degenerate constitution of the nervous system.

The nightcap patient had at times hallucinations; he was inclined to melancholy and had repeatedly manifested the intention of poisoning himself. The patient whose erethism was excited by the sight of shoe-nails had also fits of hysteria, of hypochondria, and was subject to hallucinations, and so forth. The amateur of white aprons was from his youth inclined to theft and had but a shallow understanding; he was later on subject to fits of melancholy with ideas of suicide. There were at the same time physical symptoms of degeneracy: the skull was irregularly formed and modified in a characteristic manner.

Dr. Krauss [1] mentions a similar observation, where stealing was not for a criminal object, but served merely to satisfy a perverted sexual instinct. This observation is borrowed from Eulenberg [1] and concerns a subject 45 years of age, of a hasty, spiteful character. "Often there came over him he knew not what; his head would then become heavy, hot and as if ready to burst; he could neither think nor work, and felt forced to run about like a dog. In such moments he felt an irresistible impulse to steal women's washlinen, wherever he could find any. He was never troubled by the fear of being caught; besides, he never stole other things or money. He used to put the linen on him, sometimes in the day-time, but more often at night he would lie in bed with them on. Putting them on and wearing them excited voluptuous feelings in him, and his semen was emitted involuntarily.

".... He never sold or gave away a single one of the stolen articles, but preserved them

[1] A. Krauss, Die Psychologie des Verbrechens (Psychology of Crime), Tübingen, 1884, p. 190.
[2] Eulenberg, Vierteljahresschr. f. Gerichtl. Med. (Quarterly Journal of Forensic Medicine), 1878, Bd. 28.

all in cupboards and trunks, even in his mattress and in other hiding-places. When he was arrested he wore several articles of female clothing and had a woman's shift next to his body. In his lodging were found above all in particular women's drawers and chemises, also corsets, bodices, stockings, handkerchiefs, in all 300 articles." In any case he never stole the linen of young women or girls with whom he was acquainted; he never knew to whom the linen belonged that he stole. His sexual instinct tended solely towards the female sex. Every other genesic aberration appears to have been absent: onanism, pederasty, lasciviousness in company with young lads. But the sexual desire for women, for natural sexual satisfaction, was also absent. He further admits, that for a long time past he has had no connection with women, but asserts himself to have been formerly able to accomplish copulation. Some years ago he was engaged to be married, but the match was broken off by the parents of the bride, which had greatly distressed him.

In connection with this an observation by Diez [1] may be added, relating to a boy, who

[1] E. A. Diez, Der Selbstmord (Suicide), 1838, p. 24.

felt an irresistible impulse to tear female clothing, which on each occasion was followed by emission of semen.

A very well studied example from the etiological point of view of congenital perversion of the sexual instinct is one lately furnished by Professor Lombroso, who relates the following case of a lad of 20 years of age who had a hereditary psychopathic tendency from his ancestors in the ascending line; his grandfather died insane, his mother suffered from sick headaches, he has a sister who is hysterical, a brother of his stutters, and one of his cousins is half an idiot.

With such predispositions already, the patient had also hurt his head in childhood, and it pained him for a long time afterwards; he had also from youth up been subject to intercostal pains and pains in the hips. From his third or fourth year he had likewise been subject to erections and violent sexual excitation at the sight of white objects, for instance even white walls, but particularly of linen hanging out to dry. The touch or the sound of the crumpling of linen would awaken lustful sensations in him. Ever since his ninth or tenth year he had masturbated himself at the sight of white starched linen.

Though he possessed well-developed faculties and the desire to learn, he left school at nine years of age, stole money from his parents, several times set their house on fire, was repeatedly taken up for fighting in the streets and for carrying weapons, and was finally condemned to death for murder. [1]

It is easy to understand how the further degrees of degeneracy, besides distinctly determined psychical and physical abnormities, may also present still more degraded forms of sexual perversion, for instance the impulse to martyrize his victims, to wound them, to see their blood flow, or the desire to violate little children. At the same time the lust finds complete satiety only in murdering, or disfiguring the victim, in swallowing morsels of the victim's flesh, or in accomplishing the genesic act upon the corpse.

In this connection there are two observations of the greatest interest by Demme, [2] because they present two different grades of the same sexual perversion.

The first observation is the following: Carl Bartle, a wine-merchant in Augsburg,

[1] Lombroso, Amori anomali e precoci nei pazzie. Arch. di psich. sc. pen. etc., 1883, IV, p. 17.
[2] See Krauss, Psychol. des Verbrechens (Psychology of Crime), p. 183.

37 years old, had never in his life had connection with women: on the contrary, had always felt a repulsion to them. And yet the sexual instinct was very strong in him. In his 19th year he was for the first time assailed by an invincible impulse to slightly wound young girls, from which he seemed to derive a sort of sexual gratification. He therefore slightly wounded several girls and each time had an emission of semen. But after each such act he would reproach himself for it, felt a sort of remorse and formed the resolve to conquer this impulse.

At first he limited himself to giving them little cuts, taking care not to wound the girls seriously; later he felt impelled to squeeze the arms or the throats of the girls he met. But this did not suffice to satisfy the sexual impulse; it procured him erection, it is true, but no emission. More serious woundings became necessary. He began to prick his victims with a stiletto. Curiously enough, if the girl's clothing had protected her body from the wound, he always knew that his attempt had been unsuccessful, for he then had no emission of semen.

At the same time the wounding must not be dangerous; he was too religious for that.

But he chose his victims solely among pretty girls; he spared older women altogether. He never gave himself up to onanism, although his nocturnal dreams of wounded girls led to emissions.

In his house was found a collection of finely worked stilettos, sword-canes, poniards, hunting-knives, etc., concerning which he said that for a long time past he had felt a great desire to possess such weapons. The mere sight of them, and still more the feeling of naked, smooth steel blades aroused voluptuous feelings in him which were accompanied by violent erections.

According to the information given by his acquaintances, he was a man of very quiet character, was fond of solitude and always shunned the society of females. His appearance was agreeable, and he was in comfortable circumstances.

It was found that prior to his arrest he had committed 50 attacks on young girls. Besides the stiletto, he also used as weapons lancets and embroidery needles.

The second observation of Demme [1] relates to a soldier, named Xaver, in Botzen (Tyrol), who found a special enjoyment in wounding

[1] Krauss, ibid., p. 181.

with a knife the private parts of the maidens whom he chanced to meet on his way through the streets, and then to contemplate the blood dropping from the blade of the knife; this procured him—according to what he declared in Court—the same pleasure that he might have had in actual sexual connection with his victim.

Xaver was addicted to masturbation from his childhood, was in the habit of frequenting prostitutes and had besides often committed sodomy with little girls; subsequently he found a particular pleasure in masturbating himself in front of little girls who, " in perfect innocence looked on with curiosity."

In this case he was dominated by the idea: how delightful it must be to cut the private parts of pretty young girls with a knife and view the blood slowly dropping from the blade.

According to him this "devil-inspired" lustful impulse was irrepressible, and after each attempt became more impossible to subdue, and more furious.

In this manner he had wounded seven young girls, and was not more than 30 years old when he was arrested.

According to the depositions of his comrades and superior officers, Xaver was a hot-tempered, but in no wise a bad man. He had conspicuous peculiarities, was always extremely reserved, and often remained for hours contemplating pictures of a religious character.

The following is another example of a further degree of development of sexual perversion in the same direction:

A young man, 24 years of age, entices a little girl, 12 years old, into a wood, where he first violates and then murders her, drinks her blood, cuts out her sexual parts and her heart, which last he devours. His crime having being proved, he admitted his guilt and was executed.[1]

A similar case, in which a man afflicted with cerebral disease was deliberately sentenced to death, occurred again quite in recently France. In 1880, a young man aged 19, of the name of Menesclou, was executed in Paris, for having enticed a little girl, 4 years old, to him, whom he then violated and strangled, afterwards cutting up her body into pieces.

To the grief and shame of Science the acting official experts in mental pathology, *Lasègue*, *Brouardel* and *Motet*, notwithstanding the evident and grave form of

[1] See Krafft-Ebing, Arch. f. Psych., 1877, p. 296.

psychical degeneracy of the accused, delivered an opinion unfavourable to him, declaring him to be responsible for his acts, and the patient was guillotined. When Menesclou's brain came to be examined in the anthropological laboratory, it was found that both the frontal lobes, the first and second temporal convolutions and the occipital convolution were in a state of *ramolissement* (softening). [1]

Blumröder [2] and Lombroso [3] have noted several cases of anthropophagy, in which according to the declarations of the patients the act of copulation was far from satisfying them; their lust was only entirely satiated, when they had murdered their victims and cut them to pieces, rummaged their entrails, and even eaten portions of them.

A startling example of necrophilism is that of the well-known sergeant Bertrand, a handsome young man, 25 years of age, of pleasing appearance, who used to disinter female corpses in the cemeteries, and violate them. [4]

In his confession, written in prison before

[1] See Affaire Menesclou in Annales d'Hygiène publique, 1880, p. 439.
[2] Blumröder, Ueber Lust und Schmerz in Friedreich's Magaz. f. Seelenheilkunde, 1830, II, 5.
[3] Lombroso in his study on "Verzeni e Agnolette", Rome, 1874.
[4] S. Michea, Union médicale, 1849.

going to execution, Bertrand says among other things : [1] "From my earliest youth I used to masturbate myself, without knowing what I did, and did it openly, without hiding myself. I first began to think of women when I was from 8 to 10 years old; the lust for women developed itself in me only after my 13th or 14th year. After that I knew no bounds and used to masturbate myself seven or eight times daily; the mere aspect of a female article of dress was sufficient to put me into erection. While I was in the act of masturbating, my imagination would transport me into a room crowded with women, who were all at my disposal; while I seemed to be satisfying my passion on them, I martyrized them in imagination in every possible manner according to my lust, *picturing them dead before me* and myself defiling their dead bodies. Sometimes the thought would come to me, to cut up a man's body, but that was seldom, and I then felt disgust.

" . . . . As I had not the means of procuring human corpses, I sought for the dead bodies of animals, upon which I performed similar atrocities to those which I committed

[1] Tardieu, Attentats aux mœurs, Paris, 1875, p. 114.

later on the dead bodies of men and women. I used to slit the bellies open, take out the bowels and while contemplating them masturbate myself. Then I would withdraw, feel ashamed at my conduct and promise to myself not to recommence; but my passion was stronger than my will. In doing this I experienced a feeling of voluptuousness that it is impossible for me to describe.

" .... In 1846 I could no longer satisfy myself on the dead bodies of animals—I wanted living subjects. In the camp at La Vilette, as in most other camps, there were numerous dogs around, belonging to no one and which used to follow the soldiers. I decided to entice some of these dogs outside the walls and to kill them there, which I three times did. I then pulled out their intestines, as I had previously done to dead animal bodies, and obtained the same enjoyment.

"It was first towards the end of 1846 that the thought possessed me of digging up human bodies. What first gave me the idea was the facility with which one could obtain possession of a corpse out of the common grave in the cemeteries, but I did not then accomplish it,—I was kept back by fear.

"In the beginning of 1847 my regiment

was ordered to Tours, and my detachment remained in the little town of Blere. It was here that I accomplished the first profanation of a corpse, and under the following circumstances: It was about mid-day, I was taking a walk just outside the town with a comrade, and from curiosity we entered the cemetery, which was on our way. It was towards the end of February. The previous day somebody had been buried and, in consequence of the rain that had fallen, the grave had remained undisturbed, the shovel and pickaxe had also been left there. This sight awakened within me the most unholy thoughts. I had a violent headache, my heart began to beat—I could no longer contain myself. I sought for a pretext to return at once to the town. As soon as I had got rid of my comrade, I went back to the cemetery and, without troubling myself about some labourers who were busy in a neighbouring vineyard, I seized the shovel and began with quite extraordinary strength to open out the grave. As soon as I reached the corpse, not having any sharp instrument at hand to cut it up, I set to work to hit it with the shovel as hard as I could, and with a rage that I cannot even now explain.

".... When some time after this incident I came to Douai, I again felt the desire to cut up a dead body into pieces. On the evening of the 10th of March, I went to the cemetery. It was about 9 o'clock. After tattoo at 8 o'clock no soldiers are permitted to go outside the limits of the town. I therefore had to climb over a high wall and to swim through a moat. The cold was very severe and there were flakes of ice floating about on the water. But these impediments did not keep me back. When I came to the burial-ground, I dug up the corpse of a young girl of from 15 to 17 years old. Here it was that for the first time I gave myself up to the senseless caressing of a dead body. I cannot describe my feelings; but all the enjoyment that the possession of a living woman can procure, was nothing in comparison with the pleasure I experienced. I showered kisses on every part of her body, pressed her to my heart with the phrensy of a madman; in one word, I overwhelmed her with the most passionate caresses. After having taken delight in this enjoyment for about a quarter of an hour, I proceeded to cut up the corpse and to take out the entrails, as I had done to the other victims of my madness. I then put the body back into the grave, covered it over lightly with earth and returned to barracks by the same way that I had come.

".... I was always fond of destroying. In my infancy my parents abstained from buying me anything, because I was certain to pull it to pieces. When I grew up, I could never preserve any object, a penknife for instance, more than a couple of weeks, without breaking it, and even now I still feel the same impulse to destroy. If I purchase a pipe, it is broken the same evening or at latest the following morning. It often happened, when I was still with my regiment, that on coming home a little tipsy, I smashed everything that came to my hands."

In a letter written subsequently by Bertrand to *Marshal de Calvis*:[1] ".... So far as the erotic monomania is concerned, I maintain that it did not precede the destructive impulse. It was not before the month of May, when in Douai, that I first felt impelled to violate corpses before cutting them. Up to that time I had cut up eight or ten dead bodies in Blere, without having ever felt the desire to copulate with them. I always

[1] Tardieu, loc. cit., p. 122.

acted with them, as I had previously done with the dead bodies of animals; when I had unburied them, I used to cut them up into pieces and masturbate myself in front of them. It was only after the incident of the cemetery at Douai that the erotic mania began to precede the impulse to destroy. But then the latter became more and more violent than the former, and I felt a far greater enjoyment in cutting up the bodies after violating them, than I had experienced while caressing them. In fact, there is no doubt that the destructive impulse was always more violent in me than the erotic mania. I believe, that at that time I would never have run the risk of disinterring a corpse, for the sole purpose of violating it, if I had not had the intention of cutting it to pieces. I maintain, that here the destructive instinct played the most important part, and no one is in a position to prove the contrary. I suppose I am the best judge of what was going on within me. The cutting up of the corpses was not for a moment intended to hide the profanation committed, as some have asserted: The impulse to cut up the bodies was in me incomparably more violent than the desire to violate them."

Similar sexual phrensies are observed in the last stage of degeneracy of cretinism, in idiots and imbeciles. Among these last a tendency to unnatural connection with beasts has been noticed. [1] On the other hand the extreme stages of degeneracy are sometimes accompanied by absolute brutish stupidity combined with entire disappearance of the sexual instinct.

Thus we see that congenital pederasty, like other genesic perversions, which become manifest in the field of hereditary infirmities, is not a special morbid condition, but only a symptom, and fortunately a relatively rare symptom of psychical degeneracy. Here the degree of perversion generally corresponds to the degree of degeneracy. Therefore all circumstances which lead to so-called psychical degeneracy may be dependent on some congenital disorder of the sexual activity. Here the first and foremost place is taken by hereditary disposition to nervous maladies. A father who is epileptic, or is in general suffering from any form of cerebral disease; a mother subject to hysteria, or of a pathological disposition; and, in cases of atavism, nervous disease of grand-

[1] See Mierzejewski, Forensische Gynäkologie (Forensic Gynæcology), p. 264.

father and grand-mother, or of other blood relations, are the causes which, together with other possibly hereditary morbid tendencies, are the chief factors predisposing to perversion of the sexual instinct.

After this may be cited disproportion in the age of the parents, and more particularly drunken habits in them. A fact is also worth noticing, confirmed by the observations of Flemming, Rurot, Demeaux and others, that the children of parents usually leading a sober life, are in the highest degree psychopathic, and inclined to nervous disease and insanity, when, during the act of procreation one or other of the parents chanced to be inebriated. Syphilis of the parents has also an influence the importance of which must not be undervalued among those which contribute to lower the procreating power of the organism.

At the present moment I know two families, in which both father and mother were syphilitically affected and gave birth to a number of children. The first of these suffered from the hereditary forms of the disease and were treated for it in their infancy; the others, who had been born while their parents were in the tertiary or *syphiloma* phase of the malady, exhibited during their entire childhood no ostensible signs of syphilitic affection, but they presented a perfect picture of neuropathic constitution in the field of degeneration.

One of the boys, when he was nine years old, at the sight of engravings, pictures or statues, representing naked men, used to get into a state of excitement, which would culminate in hysterical fits, and after that during several nights he could obtain no sleep. Another, eight years old, when bathing with older boys, used to become much excited, had erections and tried as if in joke to seize hold of their genital parts. On each occasion after such a bathe, the excitement, accompanied by sleeplessness, would last several days. The parents of both boys, at the time of their conception, were, as already observed, affected with syphilis in an advanced form, but showed no hereditary tendency to nervous disease.

Another cause for the development in children of a neuropathic constitution with perverted sexual sense is to be found in various depressing circumstances that may have weighed upon the parents at the time of the act of procreation. For instance, if the father or mother had but just recovered from severe illness: typhoid fever, pneumonia, physical exhaustion, a high degree of anæmia, intellectual overwork, sexual and other excesses, immoral mode of life, and so forth, in a word all that can tend to enfeeble the nervous system and the genesic force of the parent.

Lastly, among the active causes leading to psychical degeneration must further be counted the influence of climate and of soil. Among the inhabitants of high mountainous regions, the Alps, the Cordillera, the Himalayas, we find, widely prevailing, besides cretinism and idiocy, a high degree of sexual perversion or else the sexual instinct totally extinguished.

The Persians maintained that the mountains of Armenia, a high table-land between 6 to 10,000 feet, were the ancient cradle of pederasty. In this connection the accounts of several thoroughly trustworthy travellers are of great interest, who say that a prolonged residence at very high altitudes diminishes sexual desire and weakens erection, which return with renewed force on descending to the valleys.

This diminished genesic tendency might to some extent explain the relatively small increase of the population in mountainous districts, and as it is hereditary, it supplies one of the degenerating motives which impel to perversion of the sexual instinct.

Among the series of manifestations of degeneracy, the most frequently observed form of contrary genesic sense is passive pederasty, with indifference towards the female sex, and accompanied by relatively unimportant psycopathic phenomena.

Intelligence and will may be normally developed, and by this means there is always the possibility of restraining the morbidly predisposed youth from the accomplishment of the sodomistic act itself. The sexual indifference to women being congenital the inclination towards men is equally so. This sometimes manifests itself, at first quite unconsciously, in rapturous enthusiasm for manly bravery, generosity and intellectual force; later it turns to the worship of manly beauty, skill, strength, etc. Then comes the ardent desire to see the beloved object, to converse with him, to idolize him. Later on erection and emission of semen take place at mere contact with his person. All this goes to establish the predisposition to pederasty; but the propensity to the sodomistic act itself is never inborn.

When the morbidly disposed youth, under the influence of good education, has kept his mind free from stain and his imagination pure, the act of pederasty, or more correctly sodomy, itself seems to him quite as filthy and disgusting as to a healthy subject. A certain depravity, and want of will on the part of the predisposed youth, or the influence of example, of continual incitation, temptation, ruse or some degree of violence on the side of the active party, are needed to bring about the actual accomplishment of the pederastic act.

There is here nothing fatal, unavoidable, immutable. And for that very reason it is always possible by a judicious education to restrain the predisposed youth from actual sodomy, although one may not always succeed in overcoming his sexual aversion to women or his indifference towards them.

In this connection we have had numerous examples of subjects, who from their youth were aware of their perverted sexual feelings, at the same time recognizing the horror and abomination of pederasty, and the contempt with which it is stigmatized by Society—and who felt disgust at this abominable act and always kept themselves uncontaminated. In the observation of Shaw and Ferris [1] for instance, an intellectually and physically well-developed man of 35 years of age, by force of will overcame the desire to embrace a man; male society always put him into a state of erethism, and he had erection even during the medical examination. He was sometimes seized with the impetuous desire to lay his arms around and embrace some man who had pleased him; but he always restrained himself, although he was perpetually in dread that a moment would come when he would be unable to be master of himself; this dread and the wish to be freed from it, induced him to seek for medical relief.

We also find in the observations of Charcot and Magnan [2] a male subject, who from his youth felt an inclination towards men, and became so excited at the view of a virile member, that he there and then had emission of semen. Although he was quite indifferent to the female sex, and perfectly recognized the abnormality of his condition, in one word, that he was a congenital pederast, he suc-

---

[1] J. C. Shaw and N. Ferris, Perverted Sexual Instinct, Journal of Nerv. and Mental Disease, 1883, No. 2.
[2] Charcot and Magnan, *loc. cit.*

ceeded by the exercise of reason and of force of will in restraining himself all his life from any sort of pederastic practice. In this case the skilful treatment of Charcot and Magnan was not without success. The treatment adopted gave brilliant results, and at the end of a year the patient was able to have normal intercourse with women, and even to make plans for marrying.

The earlier the inborn defect is observed, the more judiciously the development of the mind and the will is conducted, and the more practical the means employed to lower and to retard as long as possible the manifestation of the sexual instinct, the greater will be the hope that the unfortunate, morbidly disposed youth may be preserved from the horrible vice. It may here be remarked that the more easily the subject learns to overcome by force of will his instinctive predilection for men, the more does his aversion to woman diminish, and in many cases of young men with congenital sexual perversion, observed by us, we found that at the age of from 25 to 30 years they became capable of having intercourse with women, of marrying and of having children.

II. Periodical Perversion of the Genesic Instinct (Periodical Pederasty).

To congenital forms of sexual perversion must also be added that morbid deviation of the genesic sense, which manifests itself at certain times, then entirely disappears, and after a certain interval shows itself again with renewed force. Such are, so to speak, temporary pederasts, who are periodically subject to an abnormal sexual propensity, fail more or less to accomplish properly the sexual act, are not seldom married men, breed children, and intermittently give way to pederasty, as dipsomaniacs give way to their lust for drink.

This periodical mode of abnormal satisfaction of the sexual instinct manifests itself most often under the form of active pederasty and of flagellation. The morbid subjects satisfy their perverted instinct two or three times during the course of the year, not oftener, and in other times have normal intercourse with women.

The more distinctly marked the periodicity of these attacks of sexual perversion, the more intense is the morbid disturbance, and the more it approximates to the form of a periodically re-appearing maniacal excitement, that is to say, the culminating exterior manifestation of psychical degeneracy.

In fact there are cases known, of intellectually gifted married men, fathers of families, who, from time to time, sometimes after long intervals, have resorted to pederasty, flagellation or necrophilism, or felt an irresistible impulse to submit themselves to the coarsest and most cynical treatment—insulting words, and blows administered by Cynedes, active pederasts or by prostitutes.

As in all forms of periodical mania, in the intervals between the attacks the patients are entirely in command of their mental capacities, and consequently are able to hide even from their nearest friends their carefully dissimulated attacks.

For instance one subject took a long time to instruct a prostitute to flagellate him in a particular manner, informing her beforehand, that after a certain time he would pay her a visit, say nothing, but without a word throw himself on the bed, and that she must then flog him according to his previously given instructions. In fact some months later he came to her, silent, gloomy, quite different to what he had been before, "so strange," as the prostitute said; he undressed, extended himself on the bed, underwent the whipping, during which he uttered some incoherent words, became much excited, had emission of semen, then slept for several hours and went away without saying a word.

After this incident he visited the woman, the fit being over, paid her the stipulated remuneration, observing that certain of his instructions had not been carried out during his attack. From this time on he used during many years to visit her at the time of his attacks, once in two or three months, and never had normal connection with her, but merely had himself flogged, and always in the same manner.

Another patient, the observation of whose case has been communicated to me by my excellent friend Dr. M. Witz, used previous to his attack to charge a person, devoted to him and acquainted with his malady, to make certain extremely complicated preparations. A particular private lodging was hired beforehand, in which were established a prostitute, as lady of the house, and with her a cook and chambermaid, also prostitutes, all three being well instructed in what they would have to do. When the attack drew near, the patient, who knew none of the persons in the house, would appear. He was then undressed, subjected to different sorts of violence connected with his sexual parts, masturbated, flagellated, etc., all in a fixed order and according to a plan previously agreed upon. He would make a pretence of resistance, swear, get into a rage, ask to be let off, but in the long run he submitted to everything.

He was then given food, ordered to go to bed, and not allowed to go out, notwithstanding his remonstrances, and he was beaten when he refused to obey. This lasted a few days. When the fit was over—which was made manifest by certain symptoms to the confidential person, who watched the whole affair without being visible—he was permitted to depart. After a few days he used to go home to his family—wife and children—who had not the least idea of his malady. As a shrewd, well-informed, rich man of business he always managed to find some plausible excuse for his absence during the period of his attack, which would occur once, sometimes twice in a year.

A third patient, who adored his wife—a

beauty—brought up his children admirably, was himself of a gentle and poetic nature, secretly seduced two boys, brothers, and gradually induced the one and then the other to become respectively passive and active pederasts. Then he used several times in the year to go with them to a bath-room and made them accomplish in his presence the pederastic act, in which he himself actively participated. At other times he felt no pederastic desires and liked the society of women, notwithstanding which he would, not without a certain amount of wit, defend the form of sexual perversion which occupies our attention.

We must also put to the account of periodical pederasty an episode of the celebrated prosecution of the "Rue Basse des Remparts" in Paris in 1845, in which 47 persons were accused of sodomy and of blackmailing.

A woman who let out furnished lodgings, who was also a procuress, deposed, that in obedience to one of her clients, who visited her from time to time and used to give her his instructions beforehand even to the smallest details, she dressed out a pederast whom she knew in her own clothes, with bonnet and veil, and a blonde wig with curls, and in that attire, which was never altered, he had to put himself at the disposition of her customer, who paid handsomely for the privilege.

To this may be added an observation of Lasègue [1] concerning periodical "Exhibitionists" (Persons indulging in *Indecent Exposure*), young, and presenting no signs of decrepitude. For instance a handsome looking, rich man, of 30 years of age, was brought before the magistrate to answer for offending public morals in churches. At the close of the day he used to enter a church, in which there were but few people, and espying a woman praying alone, would approach her and exhibit his genitals and then after a brief moment silently retire. As he declared to the physician, an irresistible impulse would seize him at certain moments, and in spite of his resistance he was obliged to give way to this fatal propensity and consciously to commit a senseless act.

In other cases known to us the fits of sexual perversion eventuated in pederasty, coinciding at the same time with connection of the subject with women, or in pederasty with flagellation, etc.

Except at the time of the fits all the

[1] Lasègue, Union médicale, 1877, 1 Mai.

above named patients were capable of normal connection with women, most of them were married and demanded of their wives no deviation from the normal accomplishment of the sexual act. Some of them suffered occasionally from terrible pangs of remorse, were overcome by deep melancholy by the consciousness of their vice, by the fear of ruining their domestic happiness or of coming into conflict with the law. But when the free interval was drawing to an end, and the fit began to approach, the patient would become uneasy, his self-assertion was augmented, and he felt an invincible longing to accomplish the sexual act in a certain perverted manner. The nearer the moment of the fit approached, the less able he found himself to accomplish normal copulation. The patient begins to fear that he would be unable to restrain himself, that he would betray himself to his wife or to his relations. Meantime the morbid desire increases, it stifles all other thoughts and desires, pursues him continually, without a moment's respite, giving him no rest by night or by day, robs him of the faculty of attending to any business, or of directing his thoughts to any other subject. He feels that if he is to continue the struggle something fearful will happen;

he fears he is going to lose all his self-command, and that he will go mad.... He gives way, satisfies his morbid lust, often looking back with loathing on what has taken place, despising himself for his weakness, and goes back to his usual manner of life and to his normal sexual activity.

As these periodical pederasts seldom confide their secret to any one, they differ by their extreme reserve from others subject to this perversion. They avoid frequenting the society of pederasts, even avoid associating with young men, do not like to converse on abnormal sexual connections, and above all betray no outward sign of the habitual pederast.

It is only illness, or the dread of being discovered, that can induce them to expose their misfortune, and in every detail given by them there can be seen the bitter self-consciousness of their weakness and moral deficiency. Many of them, in the intervals between the attacks, look with contempt upon pederasts; many indeed manifest a certain morbid hatred, particularly towards feminine looking Cynedes,—in this resembling dipsomaniacs who, in the periods between their drunken fits, cannot suffer wine at all, and cannot without disgust put up even with the smell of it.

But at the same time they have the intuition, that after a certain interval the morbid tendency will again awaken and with invincible, and irresistible force once more push them to perform a whole series of actions, the mere remembrance of which in the free intervals between the attacks inspires them with abhorrence. The dread, that during the period of their imminent attack they will not be able to dissimulate it from their friends with sufficient foresight and self-command, forces many of them, during their free intervals, to prepare for the inevitable event, and to imagine all possible precautions for keeping the matter in the greatest secrecy.

When the accustomed means of satisfying the morbid propensity has been interrupted by accidental circumstances, or the attack comes unawares upon the patient, he acts under the influence of an irresistible sensual impulse, and either employs violence, or takes so little precaution, that he draws the attention of people, gets taken up by the police, ruins his domestic happiness; he loses his social standing and in a moment of despair commits suicide, or else appears on the prisoners' bench, to the great surprise of most of his friends.

A man, hitherto considered by every one to be a modest, moral father of a family, a talented government official or a gallant general, is suddenly unmasked and discovered to be the greatest debauchee, who satisfies his sexual lust in the most unnatural and depraved manner.

Krafft-Ebing [1] mentions such a case of periodical sexual perversion, in which bestiality is concerned, but in my opinion the fact requires confirmation.

An engineer, 45 years old, father of a family, suddenly quits his business in Trieste, and hastens to Vienna to meet his wife. During the journey he leaves the train at an intermediate station, goes into the nearest village, enters a cottage, and there violates an old woman, 70 years of age, whom he sees for the first time. Arrested immediately afterwards, he declared that such an impetuous desire for copulation had been awakened within him that he had got out of the train, to seek for a knacker's yard or a slaughter-house, in order to satisfy his lust on a dog (many of which frequent these places); but

---

[1] Krafft-Ebing, Arch. f. Psychologie und Nervenkrankheiten (Magazine of Psychology and Nervous Diseases), 1877, Bd. VII, p. 296.

not having succeeded in his search and the want becoming more imperative, so that he could no longer contain himself, he went into the very first house he came to and satiated his lust on the first female he met. He had previously had several attacks of such sudden sexual desire, and had often satisfied it on dogs.

The intelligence of the delinquent was undiminished; he recognized the abominable nature of his act, explaining it by the exaggerated morbid sexual impulse which from time to time overpowered him. From his childhood he had been a neuropathic subject; from 1864 to 1867 he had suffered from recurrent mania, with exaggerated sexual desire. During the last six years he had been intellectually perfectly sane.

It is evident that we have here not only, as Krafft-Ebing says, a case of exaggerated sexual impulse, but a morbidly perverse instinct, which by chance found an uncommon way for its manifestation.

The patient, suddenly feeling the preliminary symptoms of the attack so unhappily well-known to him, hastens to Vienna, where he can secretly satisfy his passion on animals. But time presses, he tries to find a slaughter-house, around which there are always many wandering dogs, finds none, and, entirely beside himself, falls upon the first poor old woman he meets, and violates her on the spot.

If it had been simply a case of exaggerated sexual lust, he might, at least temporarily, have satisfied it by masturbation or have sought for relief in any of the houses of prostitution in Trieste; he would not then have had to run after a dog, and would not in a state of phrensy have violated an aged woman, without taking even the slightest precautions.

It is precisely this moment of phrensy that is the dread of the sufferers from periodical sexual perversion, and which induces them to prepare in time before the dreaded paroxysm.

In certain particularly marked cases the attack may fall quite unexpectedly upon the patient, as in the case of the College Registrar L., who committed the act of sodomy on a little boy two years old. [1]

L., 26 years of age, married, father of a family, gives way now and then to drink; on

[1] Mierzejewski, Forensische Gynäkologie (Forensic Gynæcology), p. 235.

the 9th of August being in a somewhat inebriated state he took the little boy Constantin from the mother and carried it on his arm to the garden, placed it on a swing and violated it. At the terrible cries of the child the mother rushed on the scene, perceived L. unbuttoned, holding the blood-stained infant on his knees. She seized the child and carried it away at once into the house. L. also immediately disappeared. The following day, on being questioned by the magistrate, to every question he replied: "That his head burned on both sides." He was the same day removed to hospital, showed symptoms of tendency of blood to the head, did not speak, breathed heavily and was low spirited. For the next three weeks he felt ill, was at times the victim of melancholy and complained of severe headache. After a first application of a cantharides blister erysipelas supervened, and it was not until the 16th of September that he came to himself again and then declared, that he knew nothing whatever of what had happened to him on the 9th August, and that it could not have been the drink, but madness that had led him into crime. It was not the first time that his senses had been so overclouded.

L. was sentenced by the Court to hard labour, but on appeal the higher jurisdiction recognized, in accordance with the official opinion of the medical experts, that the incriminated act had been accomplished under the influence of a morbid mental perturbation; he died in prison before the final decision was given.

Von Gock [1] has published the example of a still greater degree of mental disturbance and maniacal erethism, with a periodically increased tendency to pederasty.

A Jew of 22 years of age, with a weakly developed intelligence, during his stay in the establishment [2] used to seize hold of the private parts of all the employés proposing to them to accomplish with him the pederastic act, in which he desired to take the passive part. After some time his condition improved and he left the place. A few months later, however, the morbid attack recurred and with pederastic tendencies in renewed force.

In both these last cases the symptoms of a psychopathic constitution, with diminution of intelligence, were clearly defined, apart

---

[1] Archiv f. Psychologie, 1875, p. 566.
[2] Von Gock neglects to say in what establishment, probably an asylum (Transl.).

even from the attack of sexual perversion. Such subjects most frequently come to the prisoners' dock, or go straight to a lunatic asylum. But there are on the contrary other more gifted patients, more intellectually developed who, when they are convinced of having committed such a crime, seek for a refuge in suicide, like the French general N., whose affair filled all the Paris papers a few years ago.

The tendency to *necrophilism*, which is in rare cases observed in patients, in the form of isolated attacks separated by long lucid intervals, is also to be considered as periodical sexual perversion.

It is therefore extremely probable that the case of sexual perversion observed in a Church dignitary, recorded by Leo Taxil,[1] was of a periodical character. He says:

In a well-known house of prostitution in Paris there existed, according to information given by the patient himself, a chamber, the walls of which were covered with black satin on which were silver tears—a funeral decoration; at the sides of the bed were placed silver candelabra, and on it lay a prostitute, painted white all over, so as better to resemble

[1] Leo Taxil, La Prostitution contemporaine, p. 171.

a corpse, who had to remain extended, without making a movement, with her arms crossed upon her breast. At the appointed hour the prelate entered in full pontificals, knelt before the bed of the simulated corpse, muttering some incoherent words, as if he celebrating a funeral mass, and then suddenly threw himself upon his victim, whose business it was all the time to play the part of a corpse, and to lie extended without making a movement or uttering a syllable.

Brierre de Boismont [1] has recorded another observation. In a small provincial town in France a young girl, 16 years old, of a respectable family, had just died. In the night the mother of the dead girl heard the noise of furniture overturned in the chamber wherein lay the corpse. She rushed in and saw before her an unknown man stripped of all but his shirt, rising from the bed whereon lay the corpse. She raised a cry, in response to which several people hurried up, laid hold of the intruder, who seemed to pay no attention at all to what was taking place around him, and who gave only very incoherent answers to the questions that were put to him. The examination of the corpse

[1] Gazette médicale, 1849, 21 July.

showed that it had been violated, and that copulation with it had been effected several times. According to the judicial enquiry it was proved that those who were charged to sit up with the corpse had been bribed, that the prisoner was possessed of a large fortune and had received a good education, frequented the higher classes of society and had repeatedly been able, by expending large sums of money, and employing all sorts of stratagems, to obtain access to the corpses of young maidens recently deceased, which he violated. He was sentenced by the Court to imprisonment for life.

A careful and continued observation of this kind of subjects, afflicted with periodical sexual perversion, renders it perfectly possible, to discover in each of them a number of more or less distinctly marked signs of a neuropathic, excitable character or the indubitable symptoms of a heavy hereditary taint. On the other hand, a superficial and general acquaintance of such patients makes them appear as persons having very little in common with the previously described types of congenital pederasts, in their outward appearance, manners, mode of life, etc.

Recent observations would tend to show,

that the development of periodically recurrent sexual aberrations is possible without the existence of a hereditary psychopathic constitution. For instance, Anjel [1] describes the following case: A married man, of middle age, without pathological hereditary transmission, had at one time fallen in a concert-hall and considerably bruised his head, so seriously as to have remained stunned for some time. He subsequently experienced much oppression and weight in the region of the heart, etc. He was later on subject to a peculiar sort of attacks, consisting in sleeplessness, loss of appetite, irritability and mental depression. When he was in this condition the presence of little girls caused him a peculiar excitement. Even his own little daughters aged five and ten years respectively awakened within him desires, that he could only master with difficulty; the shouts of children in a neighbouring room caused him erections. He felt himself irresistibly attracted towards little girls and, although he thoroughly recognized all that was criminal and vicious in his desires, he used to go into the streets, to

[1] *Ueber eigenthümliche Anfälle-perverser Sexualerregung* (Singular Cases of perverted Sexual Stimulation), Archiv f. Psych., vol. XV, H. 2.

meet the children coming from school, and entice little girls into dark corners, where he lifted their clothes and exposed their parts.

The attack would last from 8 to 14 days; then he came to himself again, troubled with remorse, and returned to his usual mode of life, and also to normal sexual activity.

A year, sometimes 15 months, interval of tranquillity might intervene, after which the attacks recurred with the same force. Anjel considers the above case described by him as identical with epileptic fits, comparing the paroxysms of abnormal sexual desire with psychical epileptic equivalents, a point that will be considered later on. I will merely remark here that this explanation does not seem to me quite conclusive, as it is well known that the distinctive sign of epileptic psychoses consists in a certain obliteration of conscience during the attack, which, however, was wanting in the case just recorded. Psychical epilepsy manifests itself either quite suddenly, or with very brief precursory signs in the form of "aura" symptoms, and disappears with equal rapidity, which was not the case with the subject just mentioned, the paroxysm having then been preceded during several days by morbid irritability and lowness of spirits.

Again, the remembrance of the attack, which sometimes subsists immediately after it, always ultimately disappears, and this again in the case recorded by Anjel is not the case. The absence of any sort of morbid manifestation between the paroxysms, as also of the so-called epileptic symptoms, and of the obliteration of conscience immediately after the attack—all these facts justify me in reckoning Anjel's observation among the cases of periodical mania, in the group I have presented of periodical perversion of the sexual instinct. The principal interest in the above case is the absence of hereditary neuropathic predisposition.

If this fact is exact—which, however, is not altogether certified in the description—this case would constitute an exception, because in all similar observations of the kind the influence of hereditary taint is distinctly manifest.

III. Sexual Perversion of Epileptics (Epileptic Pederasty).

As Epilepsy is one of the most distinctly marked forms of psychical degeneracy, it is obvious at the outset it must often manifest itself in conjunction with a perversion of the sexual instinct. It is now every day more evident that the special study of epileptic fits, the complicated disorders comprehended under the term "Epilepsy," is far from being exhausted. In the intervals between the attacks and after the same, a whole series of supplementary phenomena may be observed, which point to a general affection of the nervous system and to a highly neuropathic constitution. In fact, a particular pathological character has been described, the so-called "epileptic crisis."

The principal points of the unamiable character of the epileptic are these: he is gloomy, now and then extremely irritable, wavering without motive from intense activity to apathy and lowness of spirits; cruel and pitiless as well as vindicative and hypocritical. All this points to the existence of a general and deeply seated lesion of the nervous centres.

Thus it may readily be understood how sexual perversion and epilepsy are often observed together, having both of them the common origin of a hereditary taint, and generally are the effects of the same etiological factors productive of psychical degeneration.

It can be asserted, without exaggeration,

that hereditary epilepsy is very often to be met with combined with abnormal sexual instincts. It often happens that the sexual power of the epileptic subject is much diminished; copulation is achieved only with difficulty, and there is not much desire for it. Sometimes epileptic subjects do not satisfy their sexual desires with women, but resort from youth upwards to masturbation. At the same time, at the age of manhood the erethism is sometimes so exaggerated, that ejaculation of semen immediately follows erection, which consequently renders normal copulation with women impossible.

The epileptic masturbators may mostly be ranked immediately after the epileptic pederasts, generally active. Besides, all the above-described forms of hereditary sexual perversion may be met with together with epilepsy.

In such cases epilepsy is simply one of the symptoms, pointing to a serious psychical degeneracy, and the manner of manifestation of the sexual perversion no longer presents any distinctive colour.

Nevertheless, during epilepsy, although in but very rare cases, peculiar forms of sexual perversion have been observed, the signification of which is equivalent to epileptic psychoses.

It is well known that sometimes epileptic patients, instead of having the real epileptic fit, are subject to sudden, temporary and rapidly evoked mental disturbances, with obliteration of conscience, mad delirious hallucinations with ideas of persecution or religious megalomania. During the course of such an epileptic disturbance, with obliteration of conscience, sexual erethism may supervene, with imperious desire to satisfy the same. The patient commits a series of nefarious acts, has sometimes abnormal sexual connection, and when the attack is over can but indistinctly remember what took place and is quite unable to explain his unreasonable actions.

Some years ago I had occasion to observe a most interesting case which, thanks to the good will of all the persons interested, did not become the subject of a judicial enquiry.

A wealthy young man, 26 years of age, had lived for about a year with a young woman, whom he seemed to love very much. During this time he had twice had epileptic fits in the night after a too free indulgence in alcoholic liquors. He led a very irregular life, but he seldom had copulation, and then generally with the same woman. He showed no tendency to perverted sexuality. One evening, after a dinner at which he had taken too much wine, he went on foot to the lodging of his mistress, spoke a few words to the maid-servant who opened the door to him, and who informed him that her mistress had not yet come home, then he went with steady steps to the bedroom, and from there into an adjoining room where a lad, 14 years old, was sleeping, whom he began to violate. The boy who was torn and one of whose hands was hurt shouted for help, and the maid-servant hastened up; whereupon he left the boy and threw himself upon the girl and violated her. He then went to bed without completely undressing himself and slept for twelve hours without awakening. When he awoke, he at first remembered nothing at all of what had taken place; a few hours afterwards he remembered having been drunk the day before and having had connection with a woman; the incident with the boy was entirely erased from his memory. I saw the patient two days later. He was unwilling to answer questions, was depressed in mind and ascribed everything to his having been drunk. A few weeks later he had renewed epileptic fits, but as far as I could ascertain no sexual perversion was this time observed.

Another case, of an epileptic subject, whose attacks were accompanied by increased sexual desire together with obliteration of conscience, has been kindly communicated by Dr. Erlicki.

Mr. X .... who had left college after a brilliant course of study, had for one or two years led a dissipated life and had some epileptic fits; he afterwards makes a journey to a country estate, and there proposes for the hand of a young lady of good family. It is settled that the marriage is to take place on the estate of the bride's parents. All the guests are assembled and are waiting the arrival of the bridegroom. He makes his appearance, accompanied by his brother, a doctor, goes straight through the saloon thronged with guests, approaches his bride and then, unbuttoning his trousers, he begins to masturbate himself before the public.

He was at once carried home and conveyed by his brother to the Hospital for Mental Diseases. During the entire journey he manifested an uncontrollable inclination to satisfy his desires by masturbation. The same tendency was remarked during the first few days of his stay at the clinical hospital but with diminishing force. After the paroxysm was over, the patient had but a dim, scanty recollection of what had occurred, much of which was entirely obliterated from his memory, and was unable to give any explanation whatever of his actions.

Dr. Kowalewski [1] (of Karkov) has communicated his observation of a case of combined epilepsy, in which during a furious maniacal epileptic attack epileptic convulsions had presented themselves, but nevertheless after they had subsided the maniacal fury still continued:

M. B...., 40 years old, previously in perfect health, was one day depressed, ate nothing, and the next morning, in presence of his wife and three children, began to importune Mme. B., a friend of the family who was sitting in the parlour, to come and copulate with him. Repulsed by her, he next appealed to his wife and, without heeding the presence of the lady friend and of the children, implored her to give him immediate sexual satisfaction. As she also refused, he fell down, began to moan, became pale in the face

[1] P. Kowalewski, Juridical-psychiatric Analyses (in Russian), 1881, p. 61.

and then had an attack of furious mania. His wife and her friend having rushed from the room, he smashed the windows, threw boiling water over all those who approached him and finally cast his wife's little three year old child into the stove.

He was acquitted by the Court on the plea of irresponsibility, and two years and a half later he entered Dr. Kowalewski's Divisional Asylum, suffering from strongly marked epileptic attacks.

Lastly, the actions of many persons accused of rape or perverted sexual acts may be imputed to epileptic obfuscation of conscience with exacerbation of the sexual instinct and impulsive acts. Unfortunately this form of transitory mental disturbance has up to the present been but very little studied.

A great sensation was created a short time ago in Russia by the case of a merchant, who, after committing all sorts of excesses in a house of prostitution in Moscow, came to himself again only in Kiev, 490 miles away, without in the least knowing how and why he got there, where he found himself alone without a farthing in his pocket. The judicial enquiry had only to do with his having been robbed, without taking into consideration his state of absence of mind. This case recalls that mentioned by Legrand du Saulle,[1] in which a French merchant, whose travelling companions had been much struck with his strange manner, found himself one day in Bombay, to his intense astonishment and alarm, instead of being in Paris.

Want of reflection, neglect of precaution, absence of any idea of avoiding the responsibility of his criminal acts, intense obliteration of self-conscientiousness during their commission, shadowy remembrance of what has occurred—such are the peculiarities which distinguish epileptic sexual perversion from the previously mentioned forms of periodical abnormality of the sexual instinct.

There now remains only to speak of a psychical condition, often alluded to by the public, but generally with the most erroneous ideas concerning the nature of the malady.

Erotomania or excitation of the psychical functions in a particular direction with an erotic tendency may spontaneously present itself in neuropathic subjects, or as a symptom of far more distinctly marked psychotic or neurotic crises.

[1] Legrand du Saulle, Étude médico-légale, p. 110.

One of the manifestations of psychical degeneration is a morbid disposition to fall in love, which is the most prominent symptom of the mental condition in question. It is more frequently met with in women, particularly among the hysterical; it is also found in men of decided neuropathic constitution, who have been addicted to masturbation, or suffer from enfeebled sexual power. Severe previous cerebral disease during youth, combined with hereditary influences, may also sometimes favour the development of this condition.

The youth, who in the company of women appears ashamed and morbidly timid, when he is alone, allows his imagination to soar at will, and has then usually recourse to masturbation; or when the erethism is still greater, together with excitability and weakness of the nervous system, his imagination alone is then sufficient to procure him emission of semen.

When repeatedly renewed influence has attached the patient to an object or to a particular person, as the subject of his love dreams, then if these favouring circumstances continue, the initial stage of this erotomania is soon developed to a more advanced form.

The patient idolizes the object of his affection, adores it, sacrifices everything to it, thinks continually of it, sends letters, composes verses and becomes unsupportable to every one about him with perpetual talk of his love, his sufferings or his ecstasy.

All observers have remarked, that the love of such subjects, notwithstanding their passion, is purely Platonic. Nevertheless absolute Platonism, at all events in the initial stages of many erotomaniacs, whom I have had occasion to observe, appears to me to be doubtful.

Most erotomaniacs are incapable of accomplishing regular normal copulation, as is the case with most of the inveterate masturbators with hereditary taint. A very frequent cause of their committing suicide is precisely this impossibility which they feel of being ever able to satisfy their flame with the object of their passion.

I have often had occasion to refer to specialists in neuropathology young men of this sort, who would threaten to commit suicide, if they could not obtain an immediate cure of their infirmity. They used to complain of incomplete erection, and that the emission of semen took place at the mere sight of the loved person, and they felt incapable of having regular sexual intercourse with their idol, without whose possession life seemed worthless to them. Not seldom their idol was only some brothel prostitute, who as for intercourse with them wanted only their money, or else some married woman who had never given the mad youth the slightest reason to suppose that she was at all inclined to allow him any intimacy whatever.

The love of such affected subjects is naturally Platonic, when its object is a person accidentally met with, or of high rank, a celebrity, etc. But here again masturbation usually procures satisfaction to their excited imagination.

As the morbid disposition becomes further developed it imperceptibly changes into disease. The patient imagines that the looks and gestures of the object of his passion, sometimes a person to whom he is quite unknown, have a particular meaning, expressing reciprocation or encouragement of his sentiments. As they are constantly exciting their imagination, and have recourse in secret to onanistic relief, they often come to entertain positive illusions and hallucinations.

The love mania becomes complicated with ambitious ideas, or else there arises a morbid dread of persecution, alternating with hypochondriacal terrors.

Congenital Cynedes, who have been brought up among women, far from the depraving influence of pederasts, but still without sufficient care and education, easily become erotomaniacs. In such cases the object of their passion is naturally either some hero, whose portrait they have often had occasion to see, or a celebrated Musician, Singer, or man of Science. The same exaggerated tendency to fall in love, but this time with perverted sexual instinct, manifests itself also in that pathological state which cannot be better designated than under the name *pederastomania*. The patient sees day and night before him the object of his adoration, whose perfections transport him to ecstasy, perfections sometimes imaginary and invariably exaggerated, to whom he swears eternal honour and devotion, promises disinterested affection, etc.

Each word, each movement of the adored object awakens in our subject either extraordinary joy, and excitation, or plunges him into despair, robs him of appetite and of sleep.

The general excitation may even increase and culminate in veritable fever,—that which Lorry has described as erotic fever, or indeed manifest itself in a regular attack of mania, with obfuscation of conscience and erotic delirium, not unfrequently derived from religious or demoniacal ideas.

We always find in the Journals and Memoirs of the great majority of Cynedes the same descriptions of sorrow and joy, of hope and fear, which characterize the correspondence and autobiography of hysterical girls and erotomaniac women.

With strongly developed erotomaniacs the object of their adoration is not always a living one, and this particularity may lead to a peculiar aberration of the sexual instinct. It is indeed well known that pictures often suffice for the excitation of onanists; similarly a picture or particularly a statue may become the object of adoration of an erotomaniac.

Ancient Greece is rich in examples of the adoration of statues and in recorded endeavours of unfortunate erotomaniacs to accomplish sexual connection with these. The story of Clysophus is well known, who was enamoured of a marble statue in the temple of Samos. He hid himself in the temple and tried to accomplish the act of love with the statue, but could not succeed on account of the coldness of the marble; he then had recourse to a piece of raw meat, placed it on a particular part of the statue and succeeded by this means in accomplishing what he wanted.

Another Greek, who had become enamoured of the statue of Cupid in the temple of Delphi, accomplished the pederastic act with it, and in gratitude for the same deposited a valuable crown at the feet of the statue. The oracle, being appealed to on this occasion, ordered that the madman be set at liberty, who in any case had already paid a heavy price for a very moderate pleasure.

But in our days too statues and pictures have at times been the objects of the adoration of psychopathics. In 1877 the French newspapers related the case of a gardener, who had fallen in love with a statue representing the Venus of Milo, exposed in a park.[1] And a few years ago, in the neighbourhood of St. Petersburg, a young man was arrested, who was in the habit of paying visits by moon-light to the statue of a nymph, situated on the terrace of a country-house.

[1] Événement, 4 May 1877.

It lies outside the object of our work, to describe rare and exceptional cases of sexual perversion, which we have not been able to observe ourselves, for instance, such as lunatics, particularly maniacs, who fancy that they are women and consequently seek for the company of men, as Dr. Raggi has lately recorded.[1] Nor is it our business to record cases of sexual mania, which is at times found in connection with sudden acts of impulse, or appears occasionally in the course of different forms of ordinary mania, alcoholic automatism,[2] etc.

With regard to alcoholism, it is to be remarked, as far as my personal observations have gone, that by itself it never leads to sexual perversion in healthily constituted subjects. Although in many cases alcohol may have an exciting effect, increases venereal desire and prolongs the duration of the sexual act, yet, as far as I have been able to observe, it does not tend to induce in normally constituted subjects any deviation from the usual method of satisfying the claims of sexual desire. But, on the contrary, in psychopathic

[1] Raggi, Aberrazione del sentimento sessuale in un maniaco ginecomasta. "La Salute." 1882, No. 11, p. 86.
[2] F. D. Crothers, Inebriate Automatism. "The Journal of Nervous and Mental Diseases," No. 2.

subjects, who are inclined to sexual perversion, and whose predisposition can only be repelled and weakened by good sense, strength of will and force of habit, the congenital tendency becomes intensified when in drink, by reason of the diminution of self-command and increased sexual desire, and the inebriate person commits a series of acts which he could always refrain from doing while he was sober.

I knew a doctor with neuropathic constitution, who usually had normal connection with women. But as soon as he had drunk wine, which very quickly took effect upon him, normal copulation no longer sufficed to satisfy his increased sexual lust. In this condition he felt impelled to prick a woman's posterior or cut it with a lancet, he required to see blood or to feel the blade penetrating the living flesh in order to find perfect satisfaction to his lust with emission of semen.

I will further mention that many authors of the present day (*e.g.* Moreau de Tours) describe under the name of Satyriasis a particular neurosis with greatly increased sexual desire, continual erection, frequent emission and morbid unconquerable desire for connection, with obscuration of the other senses.

The patient has hallucinations, delirium, falls into furious mania, becomes violent, etc.—in one word, his self-control is much diminished and he is irresponsible.

The development of satyriasis is principally attributed to sexual abstinence, particularly under the influence of religious convictions. It is for instance detailed in the confession of the Abbé de Cours, written down by himself and published by Buffon.[1] After a long period of struggle, fasting and prayer, all women began to seem to him as if surrounded with a nimbus of electric light; their aspect had a terrible effect upon him; it seemed to him as if the Governor were offering him all the Court ladies, so that he might break his vow of chastity, etc.

The apparitions which haunted Saint Anthony far surpass those mentioned in the above confession.

When patients in this state meet with opposition, they may resort to violence, and even commit murder.

Leger, who was executed in 1834 for violating and murdering a girl, had until then maintained absolute sexual abstinence, under the influence of religious convictions,

[1] Buffon, Histoire naturelle de l'Homme. Puberté.

which were strengthened by the local Minister of religion. It is easy to understand how in such a case of outburst of violent mania the accomplishment of the pederastic act is possible, even where there is a certain amount of struggling and resistance.

In my opinion satyriasis is not to be considered as an independent psychosis; it is more properly to be looked upon as a symptom of morbidly increased excitation of the cerebral processes and a general acceleration of the psychical phenomena, as is regularly the case in the maniacal state.

The sexual delirium is an incidental symptom, that shows itself in many diseases of the brain and is not at all to be taken for a particular form of "genesic mania" as Moreau de Tours maintains to be the case. Delirium of this kind may show itself in any of the above-described forms of abnormal sexual instinct, particularly if there is any hereditary taint.

Nor must we reckon as satyriasis those morbid affections which are the result of ingestion of cantharides and other so-called *pocula amatoria* (love philters). I have often had occasion to observe such patients. Continual erection, at first accompanied by voluptuous sensation and emission of semen, is succeeded by total sexual apathy, the erection continuing with bloody micturition, fever, etc.

The manifestations resulting from poisoning by cantharides should preferably be designated *acute priapism*, where also, as we shall see later on, there is continual morbid erection without voluptuous sensation.

GROUP B. ACQUIRED GENESIC PERVERSION.

I. Acquired Pederasty.

When a boy with perverted sexual instinct goes into a large school, particularly if it is a boarding-school, and comes into contact with a large number of other boys of various ages, among whom it is difficult to watch over and safeguard the regular development of puberty, he generally becomes a source of contamination to a great many of his schoolfellows.

On the one hand violent, sometimes morbidly aggravated sexual excitement, which becomes developed as the boy grows up into a youth and remains unsatisfied, on the other hand the tendency to embracing, cuddling, sleeping two in a bed,—all these render possible the first attempts at intercourse. To the above may be added habit and the spirit of imitation.

The tall, strong, active lad is always the model for the weaker and younger ones.

Under the influence of example, the wish not to be behindhand, to show their boldness, the unhappy youths conquer their repugnance to the filthy act, heat their imagination with pictures of women and at the same time indulge in pederasty. The oftener such abnormal acts are committed, the more the normal, healthy action of the sexual instinct gets blunted and modified under the influence of an acquired habit. At first a heightened effort of phantasy, with evocation of the features of a woman, was necessary to procure excitation, and to present the reality as a disagreeable but only means of relieving the exaggerated erethism. But with time the disgust gradually wears away, the reality little by little takes the place of phantasy, and produces without its aid the usual excitation. In dream as in the waking state the sexual excitation is combined by habit with the picture of the passive pederast. The image of the woman on the contrary loses its brightness, and the representation of female beauty becomes obscured. More pleasure is found with women who ape the manners of men, with closely cropped hair, breasts but slightly developed, and a narrow pelvis. After the vicious habit has become more and more confirmed, woman finally loses altogether the faculty of exciting sexually. The acquired active pederast (paedicator) becomes absolutely impotent in the presence of women, or at all events loses the faculty of accomplishing regular connection.

When once the force of example, habit, certain restrictions, etc., have developed such morbid types, the vice builds a nest for itself in the establishment; the tradition is transmitted from those who leave to the newcomers. Young, maidenish looking boys, particularly if just arriving for the first time, are subjected for their moral demoralization to a whole series of temptations and sometimes to threats and even to maltreatment, in order to induce them to become passive pederasts. To lead astray and to pervert an inexperienced lad is considered to be a sort of meritorious act, to be applauded by the pupils who have left the school. These latter generally continue connection with the establishment, visit it occasionally and on holidays invite some of the pupils to visit them at their homes, where their instruction in a certain direction is still further developed.

The school has now to a certain extent become the centre of a group of pederasts, who continue to draw thence new victims, and seek only to initiate them into the lowest paths of vice and moral turpitude. When the boy has been ruined from the sexual point of view, he is gradually taught to seek for and receive presents, treats, etc., from his teachers, that is to say to sell his charms. He thus finally becomes that most despicable representative of this disgusting vice—"a venal passive pederast."

Accustomed from his youth to masturbation, habituated for his own sake to dissimulate shame and conscience, feeling disgust for the unnatural sexual connections, but for money smothering his repugnance to the filthy act, as well as to the person of the buyer, on whom he lavishes his purchased smiles and caresses in reply to the most revolting acts that can possibly be committed on a man; accustomed from his youth to falsehood, not only in word and deed, but also looking at it as a mode of expression of sentiment connected with his profession; combining in himself at the same time all the repulsive characteristics of an onanist,—such a creature is equally venal outside of the sexual question and is capable of the most unscrupulous and dirty actions. Society has long since recognized the depravity of such individuals, and stamps these prostitute pederasts with deserved contempt.

It is in this manner that pederasty, introduced by any means into an educational establishment, causes a whole series of morbid disorders, not only in the sexual sphere, but also with regard to morality in general.

On the one hand the active pederast, who gradually loses the power of having normal connection, of his own free will disguises his nature and condemns himself to the greatest privation in life, in not only renouncing womanly love, but in exposing himself to become an object of the deepest contempt, horror and disgust to womankind; who, conscious of his vice, has not yet the force of character, to abandon it and in a fit of despair is ready to take his own life or to drown his conscience in wine and in unbridled orgies in the company of similar moral cripples. On the other hand stands the degraded, lying and venal Cynede, whose soul is fundamentally depraved, and is physically and morally ill. And between these two distinctly developed types is a whole series of perverters and perverted, a complete system of gradual allurement to vice, shamelessness and venality.

Besides boarding-schools above alluded to, sailing-vessels on long voyages, prisons, barracks with schools for soldiers' boys, etc., also furnish favourable conditions for the spread and development of acquired pederasty. But as we have already seen the combining of several conditions is requisite in order that pederasty, after it has found a resting-place, should be further developed with all its consequences. Otherwise it soon disappears as an isolated, accidental phenomenon.

The causes of the birth of acquired pederasty are not exhausted in the details given above, and in many cases it may develop itself under the influence of example and of pressure exercised by the surroundings, or else independently, as the outcome of the moral depravity of single individuals.

The intensity of the sexual impulse varies considerably according to the health of the subject. From the moment of puberty and all through life the need of sexual satisfaction, as well with regard to its force, duration and frequency, as also in the manner of its production, is in the highest degree changeable, resulting from a number of causes, which this is not the place to discuss. It is, however, important, to note that not only different persons, though quite normally constituted, but also that different races may have a sexual impulse of greatly varied intensity.

In persons of a sensual temperament the genesic function is to them, during a certain period of their life, the main-spring of their existence. They sacrifice everything to satisfy their sexual impulse, which extinguishes all other desires. Such subjects generally spare themselves no pains, are extremely enterprising and not overscrupulous about the means they may employ to attain their object, in which effort they sometimes succeed in spite of every obstacle. When through some cause or another they are deprived of the possibility of obtaining normal satisfaction, then under the intensity of their lust they have recourse to masturbation or—much more rarely—they become active pederasts. They choose the most girlish looking Cynede they can, the act is strictly limited to sodomy, and at the first opportunity pederasty is replaced by connection with women. These are so to say occasional pederasts.

But when such an individual, who has passed the greater part of his live having continual sexual intercourse with women and has never felt any interest in aught else than the sexual function, after long continued excesses, too often repeated sexual intercourse or other causes, notices that his sexual power is beginning to diminish, although the desire still persists with all its earlier force, he then has recourse to various stimulating mediums. After he has tried everything else, has heated his imagination and thereby still more excited his sensuality, whilst the genesic power continues to diminish day by day, he will sometimes resort to passive pederasty, as to an excitant to favour erection and thereby facilitate sexual satisfaction. In such cases pederasty is not the object, but merely one of the many means of excitation, which are often combined in a particular, systematically arranged method, which for a man who has been accustomed to make sexual intercourse the business of his life, and feels himself becoming more impotent every day, is now grown an absolute necessity

Debauchery may also exist even with ordinary, relatively feeble intensity of sexual impulse. The development of strength of character, the power to master one's passions and to stem them, is here of great importance.

With sensitive natures the desires as a rule quickly acquire great intensity, in that they determine certain actions or excite new desires, otherwise they give way to the dictates of good sense, of morality, habit and feeling of duty.

But when the resisting powers have not been sufficiently developed by education, the sensitiveness, inasmuch as it favours rapidly rising desires, claims for them immediate satisfaction, which in the sexual sphere finds expression in incontinence. Therefore a subject, who from the sexual point of view is feeble, may be intemperate. The slightest sexual desire awakens in him an irresistible and imperious impulse to seek to satisfy it, and this produces in a feeble subject inclined to genital weakness exhaustion, diminished excitation and imperfect erection. A peculiar intermixture is developed of impaired virility and sexual intemperance, of physical decrepitude and depravity.

The slightest excitement makes such an individual give way to his sexual lust which, in consequence of his increased sensitiveness and irritable weakness of the nervous system, grows rapidly, and is so overrated by himself, that the possibility of satisfying it turns out to be far less than he had anticipated.

Bodily fatigue and satiety come sooner than moral satisfaction. The harmony and completeness of the climax of excitation which characterize the normal sexual act are wanting. Perfect satisfaction becomes a matter of chance, and the subject, who is gradually losing his sexual power, while endeavouring to rouse the excitation, seeks to discover new means or obtains them from morbidly affected subjects who are afflicted with perverted sexual instinct. In this case imitation becomes one of the most powerful agents for propagating depravity.

As a model to imitate, there is some prominent personage, a privileged class of society, and lastly the general tendency,—the fashion of the day.

The weaker the development of the faculty to receive impressions, to assimilate them and transform them into new mental products, to reproduce them by the force of imagination and in general to exhibit, in its broadest sense, the creative power, the stronger is the tendency to imitation. And just as in questions of fashion in general an inclination to imitate points to a want of independence and to an insufficient stability of ideas, the imitative impulse gives rise in the sphere of sexual activity to a series of senseless acts provoked by excitations having no connection whatever with the sexual act.

The desire to ape a certain person, at all events to equal him in depravity, or if possible to surpass him, the longing to create a sensation or to astound by some extraordinary action, causes many petty, vain and weak-minded characters, to accustom themselves to abnormal varieties of sexual action, without the same being in any way called for by a personal want on their part. The proof of this is to be seen in the rapid propagation of various aberrations of the sexual instinct, which independently, as morbid symptoms, are very seldom met with.

There are, for instance, examples of patients afflicted with senile dementia to whom, as we shall presently see, the sight of a woman in the act of defecation caused erection. A few years ago in Paris a number of men of high standing happened to be afflicted with this morbid abnormality. At present, according to L. Taxil,[1] "les stercoraires"[2]—as they

---
[1] L. Taxil., *loc. cit.*, p. 166.
[2] Stercoraceous, *i.e.* delighting in excrement.

are named—are no longer exceptional phenomena. In houses of prostitution there are special arrangements made for this purpose, and healthy young men, from a spirit of imitation, repeat the morbid actions of imbecile subjects, who had once been famed for the dissolute life they led.

It is to be remarked that under such circumstances the spread of pederasty and of flagellation as stimulants is principally favoured by the reading of certain works, written by persons affected with congenital perversion of the sexual instinct. Such was the case of the well-known Marquis *de Sade*, the author of "La nouvelle Justine ou les malheurs de la vertu," a congenital pederast, who was several times condemned to prison and even sentenced to death for crimes against decency and for murder. At the end of his life he fell into senile dementia, and died in 1814 in the Lunatic Asylum at Charenton, where he had been incarcerated for cruel tortures inflicted on a woman. While this psychopathic author puts into the mouth of his hero the description of his own morbidly distorted feelings, he pens a warm apology for sodomy. In the images and descriptions, which are striking by their cynicism and the obvious sexual perversion they disclose, the enfeebled debauchee seeks to discover a yet untried stimulant, and he follows the counsels of a monomaniac.

The same with regard to flagellation. Le Riche de la Popelinière [1] describes in the form of unconnected dialogues and tales, rather poor in matter and literary talent, the manifestations of his own personal sexual impulse and its congenital perversion, attributing to flagellation such a stimulating power as could only be evolved in the morbid phantasies of a madman,—to which, however, the debauchee who is losing his manhood hastens to have recourse.

Pederasty also belongs to the acquired form of sexual perversion, and it prevails so to say in the endemic states among many nations in the East.

The fact is, however, that pederasty is far from being in the East so common and habitual as it is often said to be. It is there condemned by religion and to a certain extent prosecuted by the laws; but it finds notwithstanding more favourable conditions for propagation in the East than in Europe.

[1] Tableaux des mœurs du temps dans les différents âges de la vie. Avec note de *Charles Monselet*, Paris, 1867.

The absolute exclusion of women from social life, their sequestration and the impossibility for them of having any sexual intercourse whatever except in marriage, and even then often only with the permission of their parents and after the payment by the would-be husband of a sufficient sum for the possession of his bride, puts the Mussulman youths in the position of pupils of a boarding-school.

What tends yet further to develop the senile form of pederasty is early sexual enjoyment, excessive sexual intercourse in rich families, leading to satiety of the senses. Again the great disparity of age between an old man and a young woman serving him as slave is a powerful agent for producing degeneracy in the descendants, and by that means to sexual perversion.

The numerous passive pederasts, who are mostly to be met with in wealthy families, where this vice has found a traditional resting-place, do not serve exclusively for the satisfaction of perverted sexual instinct. Sometimes the wealth and consideration of a Mussulman is measured by the number of richly and peculiarly dressed boys that he has in his service. One may often see a retinue of boys, who out of vanity are kept by persons absolutely abhorrent of pederasty. Thus in the east pederasty is more openly shown, is less dissimulated, even openly vaunted, and often rich and depraved men are known to boast of possessing a beautiful boy, as with us men boast of having an expensive mistress. Putting that out of the question, the psychical interests of life in the East are more limited, and they are often more concentrated on the sexual sense, whereby depravity in its lowest form is occasioned, and may then, as we have seen, lead to acquired pederasty.

It would certainly be an error, to admit, that in the East or anywhere else a phenomenon so contrary to nature and so morbid as this abnormal mode of satisfying of the sexual instinct, which is opposed to nature and to the propagation of the species, should be recognized as normal, regular and legal.

Pederasty, like any other hideous deformity, always and everywhere rouses abhorrence in the mind of a normally developed man, and in that of woman loathing and contempt.

II. Sexual Perversion in the Dementia of Dotage (Senile Pederasty).

Sexual activity often presents many morbid aberrations during the progressive development of so-called Senile Dementia.

This denomination does not exactly correspond with the morbid state under consideration, because the phenomena mentioned are not observed exclusively in individuals of advanced age, and are far from being always accompanied by a falling off of the intelligence.

Completely developed morbid forms are mostly observed after the age of sixty, but preliminary symptoms of the same may manifest themselves at a much earlier date.

The more the individual has wasted his energy and force, either in the unequal struggle for life, or in giving way early to unbridled passions, or in the effort of surmounting serious maladies, the sooner the morbid condition appears and the more rapid will be its evolution. A peacefully conducted life, on the contrary, will exhibit in old age no morbid manifestations. Accordingly senile dementia may not unfrequently be met with in middle-aged subjects, and on the other hand men of advanced age may be found with juvenile freshness of feeling and of understanding.

The malady consists principally in a falling off in the alimentary supply to the brain, with further atrophy of the nervous elements, together with disease of the walls of the cerebral blood-vessels and consequent narrowing of their "lumen", or opening, local obstruction, dilatation and rupture of these same vessels, and so on. Whilst the above modifications interfere with the regular circulation of the blood in the brain, they also give rise to a series of consecutive morbid appearances which, besides the thickening or thinning of the cranial bones, lead to local softening of the brain, degradation and morbid weakening of the intellectual powers.

As the morbid condition in question constitutes a part only of the general lowering of the entire organism, and coincides with an abatement of the functions in the entire vegetative sphere, it finally finds expression in senile decrepitude.

This gradual decay of the organism in all the organs and systems of the body does not take place equally and may manifest itself in extremely various ways. In some cases physical weakness is the foremost symptom, with perfectly intact possession of the intellectual faculties; in others, on the contrary, there are first of all organic changes in the brain, which show themselves in changes of character and of intelligence, whereas the physical powers remain unimpaired.

In cases of the latter kind the prominent feature is often a deviation of the sexual sense, which may show itself in manifold ways. The earliest symptom, and one common to all cases, is an increasing cynicism of language, particularly when conversing with young people and even with mere boys.

Such morbidly affected subjects specially like to demoralize the latter and to deprave them, by means of alluring prints, obscene books and tales, so as to bring them in this manner first to masturbation and afterwards little by little to passive pederasty.

In other cases the sexual perversion, under the influence of hemorrhoidal accidents or of prurigo so often accompanying old age, takes another direction, and the patient himself becomes a passive pederast, while he allows his victim to take the active part.

Very often the affected subject is at one time a passive and at another an active pederast, or else he endeavours to commit the sodomitic act on little girls of four or five years old.

At first extremely cautious, distrustful and reserved, these individuals, in direct ratio with the lowering of intelligence and strength of will, and particularly of memory and understanding, lose the power of restraining themselves, allowing their abnormal condition to be more distinctly recognized, and commit brutal criminal assaults on youths and children, carrying their cynicism to its culminating point. But by reason of their continually increasing moral and physical decadence this cynicism takes at times a very naïve form, but again at other times the filthiest and most disgusting possible.

For instance, some weak-minded old man will entice children to his lodging and there expose his genital organ to them, which has long since lost all power of erection and emission.

A whole series of such cases has been described by Lasègue [1] under the title of "*Exhibitionists.*"

In one case a man, 60 years of age, holding a high position under Government, exhibited

[1] Lasègue. Les Exhibitionistes, Union Médicale, 1877, 1 May.

his private parts to a little girl, 8 years old, who lived opposite. In another, an old General, who led a very regular life, and possessed a superior understanding and education, used at times to stop before the railings in front of a house, in which there resided some little girls, exhibit his parts before them during some minutes and then silently depart. An old writer, 65 years of age, who led a quiet and respectable life, was taken up by the police for an offence against public decency, by exposing his person to women passing in the street.

All the above-mentioned patients had, at the outset of the malady, fully retained their intelligence, but later observations showed evidence of numerous pathological deviations in the sphere of the central nervous system, and towards the end of their life unmistakable symptoms of senile dementia were sure to become apparent. In other cases the satisfying of the sexual desire is accompanied by the most improbable, disgusting actions; for instance, the subject makes a woman defecate into his mouth, or himself defecates onto the naked body of a woman.

I knew such a patient who used to make a woman dressed in a very low cut ball-dress lie down on a low sofa in a brilliantly lighted drawing-room; whilst he himself, ensconced behind the door of a neighbouring darkened room, contemplated her for some time and then, being in erection, would rush in and evacuate his fæces upon her bosom, during which he experienced a sort of seminal emission. This case might be reckoned as belonging to the periodical form of perversion of the sexual instinct, had there not been concurrently a markedly increasing tendency on the part of the patient towards passive pederasty and other manifestations connected with the intellectual faculty, clearly pointing to incipient senile dementia. Still more astounding and disgusting are the morbid aberrations of those whom the French call "les renifleurs" (sniffers) and to describe whom Tardieu has borrowed the Latin tongue: "Fœdissimum tandem et singulare genus libidinosorum vivido colore exprimit appellatio 'renifleurs,' qui in secretos locos nimirum circa theatrorum posticos convenientes quo complures feminæ ad micturiendum festinant, per nares urinali odore excitati, illico se invicem polluunt." [1] (A peculiarly disgusting and

[1] Tardieu, Étude médico-légale sur les attentats aux mœurs, p. 206.

very curious type of voluptuaries is graphically expressed by the title "renifleurs",—men who gather in secret places, such for instance as the back-doors of theatres, where numbers of women come hurriedly to make water, and experience in sniffing the smell of their urine an excitement they proceed to satisfy by masturbating themselves there and then).

Under this head must also be included the cases of accomplishment of the sexual act with birds (Geese, Hens) in which the morbid state of excitement attains its climax at the sight of the dying animal, and its dying convulsions procure an extreme sensual satisfaction to the patient in the moment of coition.

Restless, irritable, sleepless, the patient's intellect sinks lower every day and he falls into second childhood. Mirth without cause and laughter alternate with tears and ill-humour; the sexual sense sometimes manifests itself in the most cruel manner. The old man thus affected will fondle a child and then suddenly chastise it without cause, and often most brutally; often his morbid lust finds excitement in the flogging of boys and girls with rods, which lead to attempts at criminal assault by seizing hold of the genital parts of the victim with his hands or in other acts of violence.

In the further progress of the malady a deep psychical perturbation shows itself, with incoherent delirium, insane notions that death is already near at hand, and dread of persecution alternating with mad excitement and megalomania. The patients often have giddiness, fainting-fits and sometimes cramps. Paralysis comes next, with complete intellectual dulness, helplessness and apathy, and the patient lets himself die of hunger, if those about him do not take sufficient care, or there supervenes senile pneumonia, disease of the bladder, etc., till eventually he becomes bed-ridden.

The disease usually takes the chronic form and pursues its course during several years. Exceptionally there are cases in which its course is more rapid, and it may terminate in a year or in a few months. The more favourable the conditions of the patient, the slower will be the progress of the disorder and the more likely are the initial symptoms to remain limited for some considerable time to isolated psychical symptoms, such for instance as despicable avarice, extreme distrust, ungrounded fear of being robbed. When at the outset the initial symptoms are of the sexual order, these subjects may become a source of considerable danger to public morals.

While they appear to enjoy good bodily health, are intellectually highly gifted, possess experience, knowledge and means, they satisfy their morbid instincts with the utmost caution and patience and proceed methodically in the work of depraving youths and children.

In view of what has preceded, one cannot chime in with metaphysical reflexions of the ingenious philosopher who could see in pederasty, particularly in the senile form, a fresh proof of the beautiful adaptability of nature. Schopenhauer,[1] it is known, was struck, previous to other observers, with the frequency of sexual perversion in old men. The extensive prevalence of pederasty in all ages and among all peoples, the impossibility of uprooting it, prove it in the opinion of this Philosopher, to be inborn nature of man, on the ground that only on such a supposition are its constant occurrence and fatal inevitableness comprehensible.

He explains the conformity of this phenom-

[1] *Schopenhauer*, Die Welt als Wille und Vorstellung (The Universe as Will and Idea), Leipzig, 1859, Bd. II, S. 641.

enon to a law by an effort of nature to avoid the propagation of the species by old men, who, as Aristotle already asserted, after their 54th year, invariably engender only weakly, rachitic offspring, causing thereby a deterioration of the race. The principal object of nature is the propagation of the species, to her the individual is only a means.

Nature manifests her laws in this respect, in that she gives to those whose power of engendering strong and healthy children is declining a taste for pederasty.

According to Schopenhauer the harm done by pederasty is but slight compared with the evil which it aids to prevent.

But actual facts contradict the paradoxical assertion of the celebrated philosopher. Pederasty, and particularly senile pederasty, causes the greatest injury to society, especially to children and youths.

Sometimes such a morbidly affected subject is for many years the pivot of a whole circle of pederasts, the head of a large company of psychopathic and dissolute men, disseminates the most unbridled licentiousness, celebrates pederastic orgies, establishes meetings for mutual flagellation, sodomy, etc., until the progressively increasing malady brings him to the grave, or senseless acts done openly render his insanity evident.

A well-defined example of senile pederasty, with slow, gradual and progressive development of senile dementia is one recorded in our forensic practice, the prosecution of Mr. J. . . .

A gentleman, 65 years of age, once occupying a very superior station and highly educated, advertised in the papers for youths to do writing-work for him, and when they came he used to enter into cynical conversations with them, counsel them to lend themselves to the most immodest actions, inciting them finally to pederasty. [1]

One of the most horrible types of senile sexual perversion, combined with the vilest cruelty, is that of the celebrated Marshal Gilles de Rayes.

A Marshal of France, Gilles de Laval Sieur de Rayes, [2] was condemned at Nantes in 1440, in the reign of Charles VII of France, for outrage and murder committed on more than 800 children during the course of

---

[1] Mierzejewski, Forensische Gynäkologie, S. 241.
[2] S. Jacob, Curiosités de l'histoire de France. Causes célèbres, Paris, 1859.

eight years, to be burned alive at the stake, which sentence was carried out.

In his castles in Brittany, where he led a solitary life far from Court, he committed the most unheard-of cruelties on young children of both sexes. When he was brought to prison, he used every means in his power for his defence, agreeing beforehand with his accomplices, as to what they must say before the Judges and what they must avoid saying. Finally, in face of the overwhelming evidence against him, and of the comprehensive avowals of his criminal aids, Henriet and Pontou, he admitted his guilt and submitted a confession, the details of which are appalling. He declared that the reading of Suetonius and the description of the orgies of Tiberius, Caracalla and other Roman emperors had first inspired him with the idea of tasting something similar himself.

Thenceforth he began with the help of Henriet and Pontou to entice little children to his castle, where he outraged them, submitting them at the same time to all sorts of torture, and then put them to death. The bodies were burned and a few pretty children's heads were alone preserved as mementos.

"I enjoyed," said the Marshal, "while

committing these acts, an inexplicable ecstasy, of course suggested by the devil, and I pray therefore that the possibility be granted me to do penance for my sins in a Cloister." He wrote to Charles VII to implore his clemency and acknowledged freely that he had left Court on account of the rising within him of an irresistible impulse to violate children, and particularly the youthful heir to the crown.

III. Genesic Perversion in Paralytic Dementia (Pederasty in the Progressive Paralysis of the Insane).

Sexual Perversion will sometimes show itself in progressive Paralysis of the Insane, and is then an early symptom of serious cerebral disease.—This affection is known to the public under the general appellation of "softening of the brain" (*Pédérastie des ramollis*).

The gradually increasing frequency of progressive paralysis, which is evidently connected with present social conditions, places this malady in the front rank of all the organic disorders of the brain, that are complicated with mental aberrations, and at once shows the necessity of taking closely into account the symptomatology of this morbid process, which is far from being always the same as well from the anatomical as from the clinical point of view.

Without going more closely into the varieties of the clinical phenomenon, everywhere known under the generic name of "paralytic idiocy", I must remark, that in the premonitory stage—contrary to the prevalent idea as to the total course of the disease not extending beyond two or three years at longest—it often drags on during a number of years before the appearance of disturbance of the motor or psychical systems.

In this prodromic period, when the changes consist principally in disorders of the vaso-motor system, there is a gradual and imperceptible change in the character, habits, inclinations and efforts of the patient, who is still considered everywhere and by everyone to be sane.

Heightened self-esteem, a somewhat greater energy, are not unfrequently in the earliest moments of the premonitory period combined with a change and a perversion of the sexual instinct.[1]

A model of a family man, hitherto temperate in sexual matters, begins to visit prostitutes,

[1] Tarnowsky, Geschlechtssinn.

candidly admitting to his physician that one woman could not suffice him. For the same reason the patient has recourse to pederasty. As he himself says the usual act of copulation does not satisfy him, he resorts to sodomy and afterwards to pederasty. He feels no remorse; he is neither conscious of the enormity of his acts, nor penitent for what he has done. He usually visits the medical man on account of some venereal disease he has contracted, and without the least hesitation communicates to him his sexual debauches and aberrations. The ease with which the patient admits the vicious satisfaction of his passion, the naïveté of the relation itself, a certain absence of shame, indifference and neglect of the universally adopted forms of description of such matters—all this must naturally appear strange to the attentive observer.

A closer examination generally reveals the exhibition of other signs, which, if not distinctly marked, point nevertheless to the initial period of progressive paralysis.

A certain absence of mind, forgetfulness and want of memory is noticeable in the patient. He is insensible to fatigue; on the contrary, according to what he himself asserts, he is always fresh, alert and in good spirits; whereas in reality he is incapable of writing a few business letters one after the other; he makes mistakes in the simplest reckonings, and easily forgets proper names. He finds it difficult to settle his mind on one subject, the intelligence in general, and the memory in particular, seem to be impaired. The patient occasionally has a tendency of blood to the head, with a dull feeling of oppression, sometimes with nausea and giddiness,—consequent upon increasing paresis of the vasomotor nerves—and he often explains all these phenomena as caused by insufficient sexual gratification, and abandons himself more than ever to all sorts of licentiousness, among others to sodomy and to pederasty.

The want of precautionary measures, which are always taken by habitual pederasts, a certain brazenness with which the affected subject will address to prostitutes or to venal Cynedes proposals of sexual connection of the most unnatural nature, as well as the absence of perseverance in the execution of his intentions, and the want of impulsiveness, which are to be observed during the periodical attacks of sexual perversion—all these signs distinguish the present form of pederasty from the other varieties of the same.

The subject importunes a woman or a youth with his filthy proposals, but if they are repelled he does not persist in his demands with the boldness and want of sense of an impulsive psychopath, but instead he makes similar proposals to some other persons. The longer the malady lasts, the more thoughtless and light-headed does the patient become.

He sometimes addresses himself to most unlikely persons, allowing himself to make use of indecent words and gestures, becomes extremely cynical, although he knows how to restrain himself on the slightest observation being made to him. In this condition he often becomes the hero of various scandalous affairs, or gives occasion to judicial prosecution for offending public decency. The malady sometimes remains very long, for months or years, in this premonitory stage, alternately augmenting or decreasing, until it attains its full development and manifests itself in the unmistakeable disorders of the psychical and motor systems, which characterize paralysis of the insane.

Ambitious monomania, irritability or maniacal excitement, difficulty of speech, tremor of the tongue, uncertain gait and other symptoms show themselves, which point undoubtedly to a deep-rooted disease of the brain.

The perturbations of the sexual instinct now fall into the background, and the patient comes under the treatment of the alienists.

At the same time it is of great importance, to recognize the malady in its very earliest stage, when it is not yet distinctly marked.

In an interesting study of the functional disturbances in the early stage of progressive paralysis Dr. Negris [1] communicates the following observation:

A gentleman, 52 years of age, who was occupied with mental labours, and led a strictly moral life, was arrested on the charge of having endeavoured to commit a criminal assault upon two little girls. Further observation showed that distinct symptoms of progressive paralysis were to be recognized in him, which had not previously been remarked.

---

[1] *Negris*, De la Dynamie ou exaltation fonctionnelle au début de la paralysie générale, 1878.

I remember a hard-working young savant, who in everything was temperate as an ascetic; with him the preliminary period of the disease lasted at least two years, during which it manifested itself, first in increased, and later in perverted sexual instinct. Whilst he was gradually losing his self-command and his intellect sinking lower and lower, he continued his pederastic intercourse with venal Cynedes, having already a chancre on his penis, and thereby infected several of them, which left him quite indifferent.

A certain blunting of the feeling of shame, of pity, a progressively increasing apathy and loss of modesty are the characteristic features of the premonitory period.

A good-natured, charitable man, who took to heart the misfortunes of his neighbour, and used to preach the gospel of self-denial and hard work, by dint of his benevolent activity and genial character, attracts the affections of a young girl, who falls passionately in love with him. They soon became more intimate and finally were betrothed. This lasted for more than a year, without any proper limits having been passed. There was a sudden break in their intercourse. The unhappy abandoned maiden suffered terribly, without exciting his compassion, and finally committed suicide with the aid of poison. His complete indifference to the tragical end of this girl who should have been so dear to him was the first distinct symptom of the commencement of a morbid process. One year later exacerbation of the genesic sense manifested itself, partly in sexual perversion (Sodomy), and three years afterwards the unfortunate subject ended his life in a Lunatic Asylum, presenting all the symptoms of regular progressive paralysis.

In order to complete my description, I will here add a few words concerning the morbid symptoms, which go by the name of *Priapism* and which, according to certain observers, should be counted among the causes which produce exaggerated or perverted sexual action.

Increased erethism, morbid sexual desire, with continual erection and emission of semen, lowered conscience and an inclination to impulsive actions, constitute, as above said, a frequent symptom of maniacal conditions of varied intensity and form.

The same may be said of Priapism or continual involuntary erection with diminished sexual desire, without voluptuous feeling, with unfrequent or long delayed emissions of semen, in many various morbid shapes.

Priapism in a particularly acute form may also be induced by intoxication with cantharides, as we have already observed.

Another similar condition appears in a relatively transitory form in certain affections of the uro-genital system (urethritis, spongitis, cavernitis, etc.). Lastly it may be one of the most unpleasant symptoms of spinal disease and of irritation of the genito-spinal centre.

For instance, in 1882, I had occasion to observe in the course of my clinical practice in the Academy of St. Petersburg a case of Priapism, of the above origin, in a soldier. The malady lasted two years and prevented the patient from doing active duty. The complete erection which was the chronic condition of the member did not diminish after repeated coition, to which the patient had at first resorted in order to be quit of his painful condition.

Later on the act of copulation and especially that of emission was accompanied by violent pain. Sexual and voluptuous desire had entirely disappeared, the thought even of copulation caused an unpleasant sensation to the patient.

In a less marked degree I have had occasion to observe Priapism, of central origin, in certain cases of the premonitory stage of tabes dorsalis (locomotor ataxy).

Priapism usually manifests itself with diminished sexual desire, involuntary long continued erection, with emission of semen, which even in copulation is very long in coming, whereby this act is extraordinarily prolonged; and even after it is ended, in the period of absolute sexual apathy, the erection still persists for some time.

The loss of voluptuous sensation, which is a characteristic of Priapism, renders it impossible to recognize in this morbid state the symptom of any deviation or perversion of the sexual instinct.

But in this there are exceptions.

As a symptom of incipient tabes Priapism may attack a man who, from the sexual point of view, has led a licentious life, considering that men given to sexual excess form an important contingent of the victims of tabes dorsalis.

The patients then ascribe the diminution of voluptuous sensation and the retardation of emission to the want of exterior excitation, to the numbing of their senses towards the accustomed charm, and they then have recourse to various new devices, hitherto unknown to them, to obtain a better erection and satisfaction of their sexual desires.

With the most unbridled orgies, Athenian nights, pederasty, sodomy, the sticking of silver needles into the erected member or into the skin of the scrotum, when erection lasts unusually long, do such affected subjects come to the end of their sexual vigour, whilst with the further progress of the malady they fall into complete impotence, combined with motor disturbances of the lower extremities and other symptoms of a rapidly progressing *tabes dorsalis*.

GROUP C. COMPLEX FORMS OF GENESIC PERVERSION.

WE have described the principal types of morbid aberrations of the genesic sense, by dividing sexual perversion into several groups, not only according to their clinical manifestations, but also according to their etiology, so far as our knowledge will allow us to do. It would, however, be a mistake to suppose that in reality such well-defined types are always to be met with as those we have represented. Between the characteristic extreme forms, there are a great many various transitions, varieties and complications, any particular type losing its characteristic features under different conditions of life, or becoming modified and mixed up with other morbid phenomena, and thus presenting itself in a complex form.

The pederast invariably seeks the company of his fellows, because it is only in their company that he can with impunity satisfy his abnormal taste, and there only meet with sympathy for his morbid condition and encouragement in his vice. Besides an active pederast recognizes more easily than does a normal person a passive pederast by his walk, by his bearing, by his gestures and his speech, his glances, etc. On his part the Cynede easily recognizes by the tone of his voice with whom he has to do. That is why pederasts in general become easily acquainted with one another and to some extent form societies, in which all the types of the foregoing sexual aberrations meet together.

Such a community excites the morbid impulses, develops them to their utmost intensity and encourages the most unbridled licentiousness. Vicious habits join to morbid predispositions, and the imagination is sharpened by the discovery of its most astounding creations, which to a sane being appear inconceivable and are fearful in their cynicism. With this the characteristics of the morbid types become blended and their peculiarities equalize each other; the Cynede learns to become an active pederast on occasion, and the latter sometimes takes the passive part.

The venal pederastic prostitutes most frequently undertake to play such a double part, but congenital pederasts also accept the same, although among the latter there is always a preference for one particular form of pederasty. The most inventive in variations of the vice, in combining it with flagellation, mutual masturbation, and so forth, are the senile pederasts, particularly if senile dementia has developed itself upon the soil of sexual perversion.

In all congenital pederasts the sexual excitability is morbidly heightened in consequence of nervously excitable weakness, so that erection soon terminates with emission and the sexual act is always rapidly brought to an end.

The desire to make the erethism last as long as possible, causes such subjects to put off the act of coition itself, and to find sensual gratification in divers contacts, and caresses, and particulary in onanizing in all sorts of ways the objects of their passion. With time this habit becomes more and more inveterate, the erethism attains its climax and culminates in a voluptuous spasm before the accomplishment of the actual pederastic act, which then becomes superfluous, and is indeed sometimes impossible, by reason of insufficient erection, or disease of the rectal tube.

In this manner the active or passive pederast is gradually transformed into a Fellator, who finds full satisfaction for his sexual lust in making use of the mouth instead of the rectal orifice: *Pompeurs de dard*, as the French call them; or as Tardieu [1] says: "*Qui labia et oscula fellatricibus blanditiis præbent.*" (Who offer lips and mouth for the fellator's pleasures).

This variety of pederasty may show itself in relatively young subjects, and is mostly brought about by example and experience in pederastic circles, which constitute the powerful centres for the propagation of moral depravity.

On the other hand the intensity of the impulse to and liking for pederasty in congenital or acquired pederasts varies extremely. Some are entirely deprived of the power of having normal connection with women; others on the contrary can accomplish it under certain conditions, for instance when the woman has a boyish appearance.

In the latter cases, by reason of repeated changing from boys to girls, the sex sometimes loses its meaning. The inclination towards a certain person and sexual excitement is then

[1] Tardieu, *loc. cit.*, p. 206.

motived by certain facial features, in which case the sex is pretty much indifferent.

A turned-up nose, a round chin with a dimple in it, sensual lips, large eyes, a warm-coloured face—make the person desirable.

When it is a woman, little developed breasts and pelvis are preferred; the stature should not be too small, and must to a certain extent resemble that of a youth.

If it is a man, he must not be too tall nor have too strongly developed muscles; taken altogether he must have a feminine appearance.

In fact, there are certain things required; the power of imagination creates a particular ideal of beauty, in which sex is of the least importance. The principal thing is the face and then a certain bodily build.

When such subjects have to do with youths, they do not always seek to accomplish the sodomistic act, and some indeed avoid it, considering that it would then rob them of the sight of the face of their darling, and they then obtain emission by friction of the member between the thighs of the Cynede, who lies on his back like a woman; the French have coined a particular term for this act: "Enfesser," and it was also known to the Romans—*Inter feces (?) coitare*, (to have intercourse between the thighs).

They avoid women in a certain measure solely because their morbidly excited erethism so much diminishes the duration of the genital act that they are unable to satisfy a healthy woman; besides, among Cynedes, particularly those so-born, the voluptuous spasm is not only induced by friction but by the mere touching of the loved person, and therefore still more by *coitus inter (?) feces* (intercourse between the thighs).

Notwithstanding all this, such subjects sometimes marry and breed children, to whom they unfortunately often bequeath in some degree their sexual aberration. They at the same time fall in love with youths, or carry on at one and the same time love intrigues with youths and maidens.

I knew a pederast, who had almost exclusively connection with youths; in a relatively advanced age he fell passionately in love with a young woman with whom he brought up several children. But he was only able to accomplish the marital act with his wife, because her features resembled those of a young man whom he had formerly loved.

Another, on the very day of his wedding

with a beautiful girl, was so passionately enamoured of a youth whom he saw, that he relinquished the proposed marriage and from that moment gave up all intercourse with women, being satisfied with *coitus inter feces.*

It is to be remarked, that this last method is the only one which certain of these subjects adopt to satisfy their lustful desires, as a variety of pederasty; but it is considered by most of those of that profession as a transitory form employed towards *coitus per anum*, particularly when leading on a youth to become a Cynede.

In this respect the seducer uses in every way the utmost circumspection, particularly acquired pederasts, in whom the impulsiveness of the act is wanting, whose power of will is not morbidly diminished; as well as in the first stages of senile pederasty, when intelligence and power are still integrally preserved.

It is only in the higher degrees of degeneracy in morbid impulsiveness, maniacal excitement, or in newly developed forms of senile dementia that there is brutal accomplishment of the pederastic act or the use of violence, without previous, preparatory manipulations.

It is also to be remarked, that sometimes the dread of syphilitic infection causes a weak degree of congenital or acquired predisposition to change into absolute pederasty.

Pederasts are generally inclined to believe, that syphilitic infection does not accompany sodomy, and that may serve to explain their attachment to this form of satisfying their sexual impulse. Needless to say, that such a pre-supposition is of course utterly mistaken.

The same forms of syphilis may be acquired in the pederastic act, as in regular coition with women. Nevertheless it often happens, that so-called accidental pederasts, after having been infected by a woman with a chancre or with urethritis, from dread of another infection give up altogether any connection with women, and become incorrigible pederasts, particularly when they find facilities for the exercise of this vice.

The history of the Roman emperors gives us the most numerous examples of utter and unbridled sexual perversion, with acquired immorality and unlimited licentiousness sprung from a hereditary basis.

Here are to be found the most complicated

forms of sexual aberration, bred under the most favourable circumstances.

Congenital predisposition, vicious education, demoralized surroundings—in one word, everything favoured the development of the most extreme, and most complicated forms of sexual aberration.

But notwithstanding all this, close examination of what has been handed down to us concerning these facts by talented and eminent contemporaries of those times, enables us to recognize in their general features the characteristic peculiarities of the principal types of sexual perversion we have above described.

From the time of Julius Cæsar to that of Diocletian we have before us a series of pathological subjects, who from the genesic point of view are extremely interesting and instructive.

Julius Cæsar was own nephew of the famous Marius, the conqueror of the Cimbri and Teutones, an habitual toper, who killed himself by drinking. Cæsar himself suffered from epilepsy, and had an exaggerated sexual instinct, which was made manifest by the enormous number of concubines kept by him.

It is well known that he wanted to have a law promulgated according to which all

S*

the women of Rome should with impunity be at his behest, so as to increase the number of offshoots of his great and glorious race.

Later, when his genesic powers diminished, he became a passive pederast, which induced Curio to call him: "*Omnium virorum mulierem et omnium mulierum virum*" (the wife of all men and husband of all women).

Tiberius is a clearly marked type of a licentious being, whose life ends in senile dementia.

At Capri cruelty little by little took the place of refined depravity, and towards the end of Cæsar's long life there were carried away thence more corpses of boys and girls martyred to death by the sick old man than flowers and odoriferous spices.

The heir to the throne, Vitellius, was brought up in Capri by Tiberius, and from his infancy had gone through a whole school-course of vice, together with many other boys, who were gradually learning to play the parts of Cynedes, beginning with the so-called "Pisciculi" (Little fishes), of whom Tiberius was so fond. (See C. Suetonius Tranquillus, De Vita Cæsarum, Tiberius XLIV).

Vitellius, an acquired habitual pederast,

after mounting the throne lived openly with the freedman Asiaticus and closed his life in a state of absolute idiocy.

The innate form of passive pederasty is most distinctly shown in Heliogabalus. Called to the Imperial throne, he made his solemn entrance into Rome clad in a semi-feminine long robe of silk, his face rouged, with painted eye-brows. He was fond of donning feminine attire, caused himself to be styled Empress and confided the conduct of public affairs to his minions whom he chose in the ranks of gladiators, wrestlers and actors. He personally dressed his lovers, had them castrated according to Oriental custom and was himself a fellator-pederast.

Active pederasty had its exponent in the person of the Emperor Hadrian, whose love intrigue with the beautiful Antinous is pretty well known. The psychopathic nature of this Emperor is best shown in the following description of his character by one of his contemporaries: "In him good alternates with evil; sometimes he is gentle, and then causelessly cruel; good-natured, or irritable and vindictive. Dissoluteness alternates with remorse; humility with manifestations of morbid self-love; justice with bestiality."

Such contradictions of character, which were so clearly defined as to strike the historians of that day, correspond entirely to the pathological products of psychical degeneration.

Hadrian was cousin to Trajan, a well-known drunkard and active pederast, on whose accession to the throne Jupiter was warned, not to let Ganymede out of his sight.

Nero—already by inheritance from his mother Agrippina a neuropathically affected subject of a pronounced type—united in himself a congenitally exaggerated sexual impulse with vicious development and at the same time a certain degree of culture, which enlarged the circle of the manifestations of his pathological activity.

After the violation of a Vestal Virgin, he castrates the youthful Sporus, dresses him in female attire and solemnly marries him. On the women with whom he copulated he exercised the most inconceivable cruelties, giving himself up at the same time as passive pederast to his freedman Doryphorus.

Towards the end of his reign he married a young eunuch, and had himself at the same time married as woman to an actor.

Without entering into further details, which

are so distinctly and truly given by *Suetonius* in his description of the twelve Roman Cæsars, by *Petronius Arbiter*, a contemporary and organizer of Nero's orgies, by *Martial, Juvenal* and other authors, we wished only to show how much the principal features of certain types of sexual perversion, even of such distant periods, and of the terrible moral depravity then existing, remain still the same. Notwithstanding the unlimited license of desire, of strong and all-powerful vice; carried to the utmost limits of demoralization; a licentiousness still further refined by the knowledge and genius of slavish talents,—in this sink of all possible sensual debauchery the pathological types retain their exact character and are remarkable for the sameness of their manner of manifestation. The omnipotent Roman Imperator shows in his genesic lust the same aberrations, as those presented in our days by an individual who has never heard speak of the Romans, nor of sexual perversion.

A Russian private soldier A. M..., released from service, devoid of all education and culture, who is indeed quite unable to clearly make out what virtue and vice signify, is brought before the criminal Court of St. Petersburg, accused of the very same actions that Tiberius allowed himself.

In the sphere of licentious demoralization neither the creative power of genius, nor the phantasy of the poet, nor the intellect of the philosopher have brought forth anything really new, anything peculiar or original, extraneously from surrounding conditions. The pathological manifestations alone bear the stamp of the really extraordinary, of relative originality, and present certain features, according to whether the sexual instinct is dulled or excited. The moral depravity of healthy subjects, with full conscience of their vice, produces the fewest new forms from the sexual point of view, and borrows its forms mostly from morbid conditions.

The refined demoralization of the Romans at the time of their decadence, may be principally attributed to the imitation of certain psychopathic rulers, these "Madmen at liberty", as the historian has called them, who entirely uncontrolled, their sense of decency obscured and their sexual lust excited, abandoned themselves openly to the most licentious acts, without being themselves conscious of their own demoralization.

The example of the upper classes carried

away the others and rendered the satisfaction of abnormal desires accessible to all. The Cynedes promenaded openly in the public places and streets and lived in particular houses. Gladiators and athletes bargained cynically with women and with men.

But leaving antiquity aside and turning to the present, we must remark that a certain indulgence towards abnormal sexual instincts favours the propagation of the vice.

This, however, cannot certainly explain the periodical breaking out here and there of fits of sexual perversion.

The first conditions for the existence of pederasty, as the result of physical degeneration, in the form of congenital perversion of the sexual instinct, or as a symptom of senile dementia, progressive paralysis, epileptic psychoses, etc., may very well have been present in all times and among all peoples. In India, in China and in Japan pederasty was well known and described many Centuries before this form of sexual perversion had spread to Persia and to Greece.

The opinion, that pederasty like an infectious disease travelled from the East to Greece and thence to Rome, and from there extended to the whole of Europe, is not quite correct.

As a morbid manifestation pederasty has been observed at all times and in all countries, and the more the social conditions of life favoured in general the development of psychical degeneracy, the greater became the propagation of abnormalities of the sexual instinct, and among these pederasty.

On the other hand it is comprehensible, that during the period of the up-springing of the intellectual activity of a nation, which displays itself in every department of art, science and industry; when by more difficult competitions there is a keener struggle for life, all leading to a higher tension of the nervous system—there are also created more favourable conditions for a neuropathic constitution on hereditary basis, with many various forms of sexual perversion.

Thus it comes that so often, side by side with a rapid progress of civilization, an increase of sexual aberration may be observed. Depravity, which formerly was almost exclusively adduced to account for sexual perversion, now combines with the rapid growth of culture, and for the same reasons, and on account of the rapid progress of intellectual development, unavoidably brings along with it an increase of mental and nervous disease.

But I repeat, that pederasty, manifesting itself in isolated cases spontaneously, attains to the height of a social evil, when it is favoured by particular circumstances, more especially by certain facilities for the satisfaction of this vicious taste. The same conditions as a rule favour the vice in its acquired form.

With us in Russia, particularly in St. Petersburg, on account of the many bathing establishments with numerous separate bathrooms with servants attached, there are a number of prostitute pederasts, who form so to say private associations.

In France, in England, in Italy the pederast always fears to find in the Cynede an informer, or at all events a black-mailer.

The procurers, who exist for such affairs, are far from being always able to guarantee the discretion of the subjects they supply.

Therefore in London, Paris, Rome and other cities pederasty requires the utmost secrecy, considerable expense and is always accompanied by a certain dread of black-mailing.

There existed besides in Paris, during a certain time, an organized supervision of male prostitution and attached to the medical police committee was a "Sous-brigade des pédérastes." [1]

In St. Petersburg the gratuity a Cynede receives is about the same as that of a prostitute; black-mailing on the part of the bath-servants is unheard of, as they do the business in partnership, and share the proceeds.

Besides the bath-servants, the Cynede contingent is recruited among young droshky-drivers, house-porters, apprentices of different trades, who have not been long in employment, etc.

Above all, the common, uneducated people in St. Petersburg, according to what all the pederasts I have known inform me, appear to be extremely indulgent to indecent solicitations—"gentlemanly games"—as they call them.

These simple people do not consider such solicitations at all insulting, and whether they accept or decline they never dream of their own initiative, to denounce them or complain to the authorities.

I knew an active pederast who for many years used to seduce young door-keepers, who at night sleep on the door mats. His advances were certainly often repulsed, sometimes in

[1] L. Taxil, La prostitution contemporaine, pp. 283, 356.

very rough fashion, but not one of them had ever threatened him with lodging a complaint.

Another used to make similar proposals with the same result while visiting vacant apartments, always making choice of young porters' assistants.

A third exploited more particularly young droshky-drivers, used to converse with them during the drive, make closer acquaintance, visited their quarters and was never subject to any unpleasantness. His proposals were either accepted or refused, but always in a good-humoured manner.

Four years ago a former soldier, Alexis M., 55 years of age, was prosecuted before the St. Petersburg district Court, accused of having subjected three lads who had been apprenticed to him, to *coitus per anum et os*. One of the victims, W. Tsch., upon whose person the judicial enquiry elicited that there were unmistakeable signs of passive pederasty, and who frankly admitted the whole scene, among other things declared as follows: "I came to Petersburg a short time ago from a village, and not knowing what might be the customs of this place, I did not complain, because I thought, that here all the masters did the same."

The common people are just as indulgent towards the proposals of congenital or senile pederasts.

In general the number of active pederasts far surpasses that of the passive.

I am of opinion that the above described facility for satisfying the vice, may partly explain the rapidly increasing number of pederasts among us, and particularly of those in whom the taste has been acquired,—a fact which has not escaped the attention of other observers. [1]

It is, however, a fact that most of these seduced subjects quit their vicious trade at the first opportunity, never again to return to it. But among the other and fortunately much smaller portion, there is developed the disgusting and thoroughly demoralized type of the Cynede prostitutes, of whom mention has been already made.

In this connection the moral downfall is the more profound in proportion as the degree of education is higher.

Youths, who on leaving school have fallen into the hands of pederasts and become Cynedes, soon got accustomed to squander their easily acquired means, and as they ascend

[1] See Krauss, *loc. cit.*, p. 176.

the ladder of depravity and crime, they abandon pederasty as a not sufficiently remunerative trade.

Others, more intelligent, combine pederasty with black-mailing,

In all great cities, London, Paris, Berlin, Vienna—great prosecutions have taken place concerning pederasty and black-mailing. "S'occuper de politique" is in the jargon of the Parisian pederasts a technical expression, which means: to be busy with black-mailing.

Children of from ten to twelve years old are, by persuasion and threats, gradually led astray to masturbation and sodomy and then trained to become denouncing Cynedes—"les petits Jésus", as they are called. Some police agent turned out of the service, or dismissed detective, who has still preserved some sort of connection with the department—creatures who have come down to the lowest grade of infamy and vice—are generally at the head of such a concern, in which prostitute Cynedes play the part of decoy-birds. [1]

It is in Russia that black-mailing is the least frequent.

---

[1] *Mémoires de Canler*, ancien chef du service de sûreté, chap. XXXIII. Les antiphysiques et les chanteurs, p. 264. Brussels, 1862.

Besides the examples furnished by Dr. W. Mierzejewski [1] the case of Mr. M... is of a certain interest, presenting as it does some analogy with similar law cases which have lately been judged abroad, although here the immediate connection of the principal party concerned with the head of the detective police remained unexplained.

The case of Mr. M. furnishes a picture of a class of society, which in Russia, besides the bath-servants, finds its representation among the prostitute Cynedes.

I shall close the description of the principal varieties of sexual perversion by a short extract from the hearing of this cause, which gives us the lowest, and most infamous expression of pederastic prostitution and black-mailing.

In the commencement of the seventies a personage occupying a high situation in the administrative world of St. Petersburg was accused of pederasty, and immediately without the case being heard or pleaded in Court, he was removed from the government service and exiled abroad.

The denunciator was a mercenary pederast,

[1] Forensische Gynäkologie (in Russian), Case 36 & 37, p. 252.

the son of a coachman, rather weak-headed, but a shameless young fellow, without trade or education.

A great deal was said about this affair. The victim was married, had children, was esteemed to be a model father of a family, and distinguished for intellect and culture; he had before him a brilliant career and had lost all through the denunciation of an unknown vagabond.

This event did not fail to have a particular effect upon the mercenary pederasts. They perceived that it was sufficient therefore, on any favourable occasion, to threaten any government official with an accusation of pederasty, and from fear of losing his situation, he would no doubt compromise the matter and cash up in order to stifle the accusation. The most infamous black-mailing prevailed.

The principal hero of this was a certain youth of 17 years of age who had quitted the second form of the Gymnasium (High School) and resided with his mother—a poor woman who just managed to live thanks to a monthly stipend of 25 roubles allowed her by a relation.

In the meantime M... paraded about in

the hotels and restaurants, which he assiduously frequented, known under the name of Mitroschka, dressed in a crimson silk shirt with a blue sash, over which he wore an overcoat. He squandered money in droshkys and at billiards in the cafés. He was seen to have about him as much as from 200 to 300 roubles.

His accomplices, some of whom appeared on the prisoners' bench together, the cases against the others being dismissed by the Court for want of evidence, were young fellows without means or occupation.

One of them, Pr..., 17 years of age, was the son of a retired Sergeant, who carried on a small commission agency, and with whom he resided; another, aged 19, had first been a singer and then a servant, but had lost his place. A third, 25 years old, was waiter in a hotel; the fourth was a clerk out of work; the fifth, aged 21, a tailor without occupation, and so forth.

All these young fellows met together daily in a well-known restaurant, where their persons as well as their nicknames were known to the waiters. There acquaintances were made, particularly with travellers newly arrived and rendezvous given.

Generally M. used to hire a bath-room

along with the intended victim; after a little time one of his accomplices would rush in, make a noise, threaten with an accusation of pederasty, with denunciation to the secret police, and finally allow himself to be pacified in consideration of a certain sum of hush-money, "not to proceed further in the matter." The rascals shared the money together, and the victim naturally held his tongue for fear of raising the slightest suspicion of his disgusting vice.

The thing was brought to light, by a case in which theft was added to black-mailing.

Under the circumstances detailed above M. stole from Mr. J..., an official just arrived in St. Petersburg, a portemonnaie containing cash, and visiting-cards, besides his watch and chain; two days later M. dined with his accomplices at one of the best restaurants and deposited in guarantee for the payment of the expense of the dinner the visiting-card of Mr. J., also signing in his name the bill claimed by the keeper of the restaurant.

When Mr. J. was applied to at his address for payment of the above account, he went to the Minister of Police to complain of the robbery of which he was the victim, fearing also a further misuse of his name. A judicial enquiry was instituted and M. was taken into custody. He began by asserting that Mr. J. had endeavoured to allure him to sodomy, and mentioned the names of some other persons who, according to him, were also given to pederasty.

A few days later the same M... also accused to the secret police another official, Mr. B., just arrived, with having tempted him to commit sodomy. Shortly afterwards another complaint was lodged with the minister of police by an official, Mr. E..., to whose lodging two young men had come, pretending to be agents of the secret police and who exacted money from him, threatening should he not comply with their demand, to accuse him of pederasty. Mr. E. gave them some money, but anticipating that the attempt might be renewed he communicated with the police.

It was proved that M. was one of the young men who had extorded money from Mr. E.

But the effrontery of these black-mailers was still greater at the expense of another government official, B. They went to his office demanding to be paid a sum of money, failing which they threatened to denounce him as a pederast. He was several times weak enough to yield to their threats and gave them altogether 25 roubles. After that they continually presented themselves in the lobby of his office, and their demands exceeded all bounds.

When orders had been given to refuse them admittance to the office, they went to the private residence of Mr. B., and called for his brother, to whom they declared that unless 50 roubles were immediately paid to them they would lodge a complaint of pederasty against Mr. B.

Finally M. was found guilty of robbery to the amount of less than 300 roubles and of obtaining money under false pretences and threats, and sentenced to six months' imprisonment.

Up to recent times forensic medicine massed together all the forms of pederasty under the common denomination of sodomy, without even attributing any particular importance to the separation of the latter into active and passive, taking it for granted that both always occur together. Exaggerated sexual lust, demoralization, satiety of lustful desire—these were considered by most medical men to be the cause of pederasty.

How difficult it is for the medical jurist to free himself from this mode of thinking, considering that prosecutions for pederasty mostly present a mass of extremely immoral acts, may best be seen in the works of Casper and Tardieu, whose studies on this subject have hitherto been considered classical.

Casper relates for instance the following episode:

The house-porter F. committed onanistic acts in the most abominable manner on five children, but at the same time not masturbating himself. The cranium of the accused was remarkable for its resemblance to that of an ape: the forehead was quite flat, the cheek bones and upper maxilla stood out prominently.

Two months later a school-master F. was charged with similar onanistic acts on two boys and three girls. He also had an extraordinarily shaped head: very prominent cheek bones and upper maxilla with the posterior part of the cranium arched.

This formation was so remarkable, that it led to my being consulted as to whether it might bear upon the degree of his guilt. I drew attention to the simian form of the cranium of the recently condemned F., which so closely resembled this one, but denied the necessity of drawing any conclusion therefrom. The offender was sentenced to a long term of several years' imprisonment.¹

Tardieu, who has written a monograph extremely rich in facts on pederasty, says nevertheless at the end of his work: "however incomprehensible, however contrary to nature pederastic acts may appear, they cannot escape either the responsibility of conscience, the just severity of the law, or above all the contempt of decent folks."²

On looking more closely into the matter it is impossible to expect remorse of conscience from a congenital Cynede, who from the first moment of the manifestation of his sexual instinct has felt and known no other impulse but that of pederasty. It would certainly appear to be no less unjust to inflict the full penalty of the law on a morbid subject in whom the first symptoms of a long and painful disease manifest themselves in pederastic attempts.

So finally, if we nowadays look at im-

---

¹ Casper, Praktisches Handbuch der gerichtlichen Medicin (Practical Manual of Forensic Medicine), Berlin, 1880, p. 199.
² Tardieu, loc. cit., p. 206.

pulsive acts as the result not so much of a wicked as of a diseased will, it is evident an epileptic pederast must rather inspire us with sorrow and commiseration than with contempt, as is the case with subjects afflicted with dipsomania or other forms of mental aberration.

For the psychopathic child with morbid sexual instinct, it is not so much the punishment of vice that is required, but rather a proper education and treatment. In a far greater degree does the individual attacked with initial progressive paralysis, or senile dementia, require medical aid, instead of punishment.

In all these forms it is either possible to alleviate, even to cure the morbid condition, or else, by isolating the patient, to render him harmless to society; in this case a penalty for immoral actions is not applicable.

Such subjects are of neuropathic nature, the victims of mental or nervous disease; they are *not* criminals.

From a number of examples of manifestly diseased individuals who have been condemned, I will mention the following cases which occurred in France a few years ago.

A man named R.... murdered an old

woman, 53 years old, and violated her dead body. He then threw the corpse into a river, but a short time afterwards fished it up again, to renew carnal connection with it. R. was condemned and executed. Dr. Evrard, who made the post mortem examination of his body, found numerous morbid alterations of the brain and of the meninges, for instance a notable thickening and adhesion of the meninges to the frontal convolutions, etc.

Dr. Cornil, who communicates this fact, adds the following judicious remark: "If the Bench of Judges consider the guillotine to be a curative method in the treatment of the insane, the fact should be made generally known."

In the case of Menesclou, to which we have previously alluded, a manifestly morbid subject was guillotined owing to the gross error of the medical jurists—an error which was proved by the post mortem examination of the brain of the unfortunate man.

It is only real vice, the acquired sexual perversion of a healthy man, particularly the vice which is represented by prostitute pederasts, that can justly deserve punishment. And here it may be asked in the name of Justice, if the same degree of penalty is to be meted out to a weak-minded, simple country lad, accidentally led astray as to thoroughly depraved and utterly demoralized mercenary habitual Cynedes?

However that may be—punishment has a logical meaning in cases of acquired pederasty.

Consequently at the present time the technical examination concerning pederasty is very complicated.

The expert has not only to decide whether the subject is a pederast or not, but also to determine, what form of sexual perversion he has before him.

I will here devote a few words to this subject.

First of all, the question is, whether it be possible to recognize with perfect certainty a pederast *by outward signs?*

With regard to the passive form we may here answer in the affirmative.

It is undoubtedly easiest to distinguish the Cynedes from among the various kinds of sexual perversion, by the deformations which are observable in the *orificium ani* and the neighbouring parts.

These deformations, taken singly, are not particularly characteristic, but taken all together, they give a sharply defined picture, which after a little practice makes the diagnosis easy.

In the text-books, and even in such a valuable monograph as Tardieu's work, the signs of sodomy are described singly, and a description of the general appearance of the alterations noticeable on what may be called an average pederast is wanting.

This circumstance has caused some less experienced enquirers in this branch of research to deny the existence of unmistakeable signs. So, for instance, Casper and Brouardel deny the importance of the infundibuliform (funnel-shaped) widening out of the anus, which Tardieu so particularly insisted upon.

This fact must also be considered, that the conditions, under which the medical Jurist and the clinical Physician conduct their examination, are widely different.

In presence of the first efforts are in most cases made to dissimulate existing changes, or to simulate those which are wanting; but before the latter nothing is hidden or pretended.

This explains to me why, as we shall see further on, certain modifications which to the practitioner appear extremely remarkable, are not sufficiently taken into account by medico-legal experts.

The last fact may be partly accounted for in that the medical Jurist cannot possess the same practical skill in the examination of the *orificium ani*, as can be attained by the specialist, who has to see daily some dozens of patients and who gives special attention to the rectum, so often the seat of syphilitic affections.

Besides, the medical Jurist has seldom, and only at distant intervals, one or two Cynedes to examine, whilst the general practitioner is obliged on the same day to examine a whole series of such individuals.

As an instance I may mention that during the course of last winter, being summoned together with my colleague Dr. Seweke to examine the pupils of an educational establishment, where a syphilitic contagion had broken out, we had to examine in one day 29 passive pederasts of from the ages of 9 to 15, among whom we found 23 presenting the most unmistakeable signs of sodomy.

The carefully noted and repeatedly controlled results of such examinations authorize me to modify to some extent the evidences of sodomy as given in different works and text-books.

We will begin with the *passive habitual pederasts*.

I will first of all describe the entire picture presented by the deformations noted in Cynedes of from 10 to 16 years of age; I will then pass to the description of each particular symptom.

The boy subjected to examination must be placed kneeling across a broad bed, with his breast reclining on a pillow, so that his head lies rather lower than his posterior which must be thrust forward; the legs must be drawn asunder, so as to be as wide apart as possible.

Placed in this position, the person to be examined, if he is ignorant of the object of the exploration and has no intention of hiding anything or of simulating, will distinctly show the signs of habitual pederasty.

When the legs are sufficiently separated, they cannot in this position be in contact with the buttocks and the anus is clearly disclosed to view.

The anus then no longer presents the appearance of two folds of skin united on the same plane by the sphincter, but rather an infundibuliform fossa (funnel-shaped depression), the walls of which, beginning from the outer lower border of the sphincter, continue funnel-shaped downwards, gradually narrowing towards the posterior depth, commencing from the upper contracted layers of the sphincter, whilst above and exteriorly they change into that form of the epidermis surrounding the anus, which in the normal condition forms radiary folds around the anal orifice. The latter are obliterated, and consequently the transition from the border of the sphincter to the inner surface of the buttocks fails to exhibit that change from the radial form of folds which is noticeable in the normal condition.

When the Cynede has been placed in the above-described position, and the buttocks have been a little opened out by exercising a pressure with the thumbs of both hands, the funnel-shaped orifice of the anus is correspondingly opened out, and the walls of the rectum become visible.

This widening of the orificium, the result of a considerable relaxation of not only the outer but also of the inner coats of the sphincter, is in my opinion the most characteristic testimony.

It may not always be noticeable in some Cynedes, but when it *is* present, the subject is undoubtedly an habitual passive pederast.

Further, the exploration of the rectum with

the finger shows a very considerable relaxation of the sphincter ani muscle.

The finger introduced into the rectum is no longer held tight by the sphincter, but in fact two fingers may be easily introduced.

It often occurs that exploration with the fore-finger may cause pain because of slight lacerations of the border of the orificium ani, where the exterior epidermis blends with the mucous membrane of the rectum.

When therefore, the subject under examination being placed in the position as above, a funnel-shaped cavity in the anus is observed, the radial folds are obliterated, separation of the buttocks easily causes a widening and gaping of the orificium ani, and introduction of the finger into the rectum does not cause a contraction of the sphincter, we have undoubtedly a Cynede before us; more particularly if it is proved that previous to the exploration the individual has never undergone any surgical operation on the anus or the rectum.

But the characteristic feature may easily be obliterated by various circumstances, some of which we shall briefly mention.

The Cynede, in order to dissimulate his condition squeezes the nates together and thus contracts the sphincter and the levator ani muscle. Therefore, in such cases, when it is required to make the orificium ani distinctly visible, the nates must be forcibly separated, when the cavity produced by the drawing in of the anus resulting from the contraction of the levator ani muscle becomes apparent. Such an appearance may also show itself in normal subjects, when there is a vigorous contraction of the muscles combined with a forcible thrusting asunder of the nates. I often caused notorious Cynedes to draw together the nates and contract the muscles of the anus, by pressing their legs together. In this position, when the nates were forcibly held apart, the characteristic alterations of the anus due to relaxation of the muscles were no longer visible. The exploration may therefore give quite different results according to whether the subject under examination has contracted the muscles of the anus or not.

In order to avoid mistakes in this matter, I took care to observe how long a 16 years old Cynede could continue the contracture of the muscles of the sphincter and of the anus, while in the above kneeling position with the legs spread apart.

I noticed, that after 10 minutes, notwithstanding the will of the subject under examination, a temporary relaxing of the contracted muscles took place, and at this moment the characteristic alterations became plainly visible.

At the end of 15 minutes the nates drew asunder and it was only now and then that the orificium ani was slightly drawn in and again lowered by a contraction of the levator muscle, because at the moment of this lowering a slight opening out of the nates caused a characteristic gaping of the anus. Every change of position of the subject under examination prolongs the affair, because the tired muscles find time for recuperation, and therefore it seems to me—in contradiction with Tardieu's opinion—easier to tire out the Cynede, by keeping him in the same kneeling position from 10 to 15 minutes, than by continually shifting the position of the seat of exploration, as Tardieu recommends.

At the same time the above-mentioned observation of the wilfully contracted and then relaxed anal muscles of the Cynede, clearly disproves the lately expressed opinion of Professor Brouardel concerning the formation of the anus infundibuliformis. [1]

[1] Brouardel, Étude critique sur la valeur des signes attribués à la pédérastie. Annales d'Hygiène publique, 1880, p. 182.

The funnel-shaped depression of the anus in Cynedes, to which Cullerrier has already drawn attention, and which Tardieu considered as the most convincing proof of passive pederasty, cannot be explained solely by the contraction of the levator ani muscle, as Brouardel supposes.

It is easy to be convinced of this, by observing the characteristic infundibuliform anal opening in a Cynede, when both the sphincter and the levator ani muscles are relaxed.

On the contrary the contraction of the above-named muscles invariably diminishes the peculiarity of the existing well defined funnel-shaped hollow of the anus, which, when the nates are then forcibly separated, is transformed into a slit shaped anal opening, like that which may be seen on every normal being, but which Brouardel erroneously took to be a typical infundibuliform depression.

The anus infundibuliformis is not at all caused by a contraction of the levator muscle, but solely by a deformation in the sphincter. When a large virile member penetrates into the opening of the rectum the lower and weaker portions of the sphincter give way more easily to the pressure, whereas the upper and stronger, contract with energy and to some extent defend the entry into the rectum itself. Consequently the penis when introduced pushes the superior muscular layers upwards at the bottom of the rectum, thrusting asunder at the same time the lower portions of the sphincter muscles some 4 centimetres in thickness, and meets with greater resistance on coming to the upper muscular layers. When the act has been several times repeated the lower portion of the muscle is widened out, and then constitutes the basis of an infundibulum which is limited by the border of the nates, whilst the upper muscular layers are thrust back and upwards in the form of a slender ring, which closes the entry into the rectum and forms the apex of the funnel.

The process by which the anus infundibuliformis is formed, is precisely similar, as Martineau justly observes,[1] with that of the development of a similar depression in the exterior genital parts of little girls who have been subjected to repeated attempted criminal assaults, or on those of adult women who have had to do with a membrum virile of unusual size.

[1] Martineau, Leçons sur les déformations vulvaires et anales, etc., Paris, 1884.

The infundibulum vulvæ is formed at the expense of the musculus constrictor cunni, the same as the infundibulum ani is exclusively formed by repeated pressing of the sphincter.

The more gradually the Cynede has become habituated to the sodomic act, and the greater the size of the membrum virile introduced, the larger will be the funnel-shaped hollow, the basis of which is continually being extended round the orificium ani, at the expense of the radial skin-folds which are gradually obliterated.

But—I repeat—the above-described symptom is valuable, if it becomes clearly visible, without there being any forcible thrusting asunder of the nates, or even when there is but a slight separation of the same. Whenever it is necessary to forcibly separate the nates, in order to expose the anus to view, this symptom loses its importance. If, on a normally developed individual who purposely contracts the orificium ani, the nates are separated with gradually increasing force, there will always come a certain moment, when the inferior layers of the sphincter become expanded and the entry to the rectum is closed solely by the upper layers. At this moment the orificium ani forms a long and deep slit, very much resembling the anus infundibuliformis.

It was precisely this confusion of the infundibulum when the muscles of the anus are relaxed, with the slit-shaped hollow when the nates are thrust asunder and the sphincter and levator ani muscles are contracted, which induced Brouardel to question the diagnostical signification of the anus infundibuliformis and to maintain, that the same symptom might appear in cases of irritation of the anus through cold, painful fissures, inflamed hemorrhoidal nodes, etc.

Further, it must be taken into account that a well-developed anus infundibuliformis, as we have already observed, loses its specific character and changes into a slit-shaped hollow, when the nates are forcibly separated and there is contraction of the muscles of the anus.

Therefore the anus infundibuliformis may quite escape observation on very corpulent Cynedes with highly developed nates closely pressing together; and certainly its absence is not a sufficient proof against the existence of pederasty.

It may also occur on very lean subjects,

where the orificium ani is almost on the same level as the thinly developed nates, where the thickness of the sphincter muscle is itself very slight, and even in habitual Cynedes the funnel-shaped hollow is but slightly and indistinctly defined. Similarly, when the membrum virile is of but small size, and although its intromission may have frequently taken place, the infundibuliform hollow may be absent in the anus of an adult subject.

Finally, in senile passive pederasts the infundibuliform hollow may also sometimes be quite indistinct, by reason of the numerous stages of the development of hemorrhoidal nodes, and is sometimes by prolapsus ani rendered so difficult to discern, as to make it impossible to found a diagnosis upon this symptom.

So it appears from the above, that in the diagnosis of sodomy the anus infundibuliformis can have but a relative, and not a decisive value, as Tardieu supposed.

Another, so to say, classical sign of sodomy, is the obliteration around the anus of the radial skin-folds—"Podice laevi" (with *smooth* fundament) of the Roman Satirists, a symptom, upon which Zacchias in the XVIIth century insisted, the importance of which was not denied by the sceptical Casper and which Tardieu highly prized.

There is no doubt, that every practising physician can settle the question of Casper's doubt about the origin of the above symptom and maintain with certainty, that it is not the smearing of the anus with the fatty substances employed by pederasts that can bring about the said deformation, but the repeated forcible expansion of the anus and surrounding epidermis by the sodomitic act itself.

People afflicted with habitual constipation are used to daily anal inunction of fatty substances, and during a long period of years are continually taking clysters, and also others afflicted with eczemas in the neighbourhood of the anus. But notwithstanding, in such subjects the radial folds around the anus are not obliterated, because the inunction alone and the careful introduction of a clyster nozzle cannot produce a stretching of the skin surrounding the anus.

Notwithstanding all this, the absence of the radial folds is in itself, in my opinion, but of very slight importance.

I have often observed on muscular, well-developed youths and men with fleshy buttocks, that these being held asunder, the skin-folds were absent, and yet these subjects had never been addicted to pederasty.

On the contrary I have found the radial folds on notorious Cynedes, together with well-developed infundibulum ani. Out of 23 Cynedes, whom I examined together with Dr. Seweke a short time ago, and who presented the most unmistakeable signs of passive pederasty, there were only 12 on whom the radial folds were quite obliterated, whereas on the other 11 they were distinctly visible.

A far more important element in the diagnosis of passive pederasty is undoubtedly the atony of the sphincter muscle, which is perceptible when the fore-finger is introduced into the rectum.

When the sodomic act has been completely accomplished and the intromissio penis into the rectum has been repeatedly effected, the first in order of symptoms to present itself is relaxation of the sphincter. But the use of this symptom, taken by itself, and in not very inveterate cases, requires a certain amount of practice. Having daily to do with a great number of patients suffering from affections of the genito-urinal organs, whose rectum I examine, I attach considerable importance to the pressure exercised by the sphincter upon the finger introduced.

It is probable that the absence of these conditions, which make it possible to compare passive pederasts with normal individuals, may explain why the majority of medical Jurists, and among them Tardieu, do not speak of the results of the manual examination of passive pederasts.

When a Cynede spontaneously submits himself to examination, seeks medical aid on his own account and no longer sees in the physician a representative of legal force, when he has no cause to hide anything or to simulate, the introduction, in the above-described kneeling position, of the finger smeared with vaseline into the rectum is easy, free and painless. The digit glides imperceptibly through the sphincter and penetrates into the profundity of the rectum, being as little gripped by the muscle as by the walls of the intestine.

The feeling is exactly similar to that experienced when manually exploring the vagina of a young girl who has been deflowered a few months previously.

Further manipulation, such as twisting the intromitted finger round, feeling the prostates, withdrawal of the digit from the rectum and its re-introduction therein, do not either provoke any pressure on the part of the sphincter.

When the atony is still more pronounced after exploring with one finger two may also be introduced with equal facility.

It is quite another thing when the individual to be examined per rectum, is a perfectly healthy subject, and particularly if a youth or a child. In this case the finger introduced is clasped by the sphincter as by an elastic ring, and each movement of the finger—further introduction, turning the finger round, feeling the prostates, etc.—will be accompanied by a fresh shock due to the involuntary contraction of the muscle under the influence of the unaccustomed irritation.

As already remarked, this action is more distinctly observable in youths and children. The older the individual becomes, the less in general is the contraction of the sphincter.

The above described symptom naturally loses its diagnostic value when it applies to subjects over 40 years of age, and more particularly to old men, who are subject to hemorrhoids, or have had to undergo an operation for artificially widening the sphincter.

When the atony of the sphincter is more distinctly marked, it then gives rise to the appearance of another symptom, which in my opinion is far more characteristic as compared with the others.

I allude to the gaping of the orificium ani, the result of which is to open out to view the walls of the rectum to a depth of several centimetres. This opening out of the upper covering layers of the sphincter orifice may occur involuntarily, as soon as the Cynede about to be examined takes the kneeling position, reclines upon his breast, lowers his head, lifts up his loins and thrusts the anus forward.

This spontaneous gaping of the anal orifice generally lasts a few moments, until the gradually contracting muscles close the opening. But it is sufficient merely to slightly separate the nates, and the orificium ani gapes again.

Finally, when old Cynedes are placed in the kneeling posture, with the legs held apart, whatever effort they may make, the sphincter is no longer able to contract sufficiently to close the orifice, which during the whole time of examination remains open, gaping.

The above symptom is so far of great value, that it requires from the explorer neither special practice, nor much experience, and can be proved even on such Cynedes as endeavour to dissimulate their condition, care being taken to thrust them suddenly down without warning into the kneeling posture, with head bowed down. In the first moments, until the subject under examination can succeed in obtaining a contraction of the sphincter while in this position, the orificium remains wide open. This gaping of the orificium ani can also be observed even after operations performed on the rectum, by which the sphincter has been sectioned or artificially dilated to its utmost extent. It may also sometimes be seen on very decrepit old men, or on very emaciated young people, for instance consumptive patients or convalescents after dysentery, typhus, etc.

The possibility, when there is an involuntary gaping of the orificium ani, or when the nates are held asunder, of seeing to a certain depth the walls of the rectum, furnishes also in passive pederasts another symptom hitherto but little noted.

I have in fact often observed in the borders of the upper layers of the sphincter and of the intestinal walls, lacerations and fissures extending lengthways. On forcibly separating the nates and strongly pressing the sphincter these surface fissures sometimes yielded a few drops of blood.

The formation of such fissures which cause but little pain or itching and readily heal, in persons affected neither with syphilis nor with hemorrhoids, always indicates that they owe their origin to passive pederasty, when there exists at the same time a distinctly marked relaxation of the sphincter.

With regard to other morbid manifestations, such as for instance the formation of abscesses in the cellular tissue around the anus, of fistulas, hemorrhoidal nodes, verrucose excrescences, etc., their presence or absence is of no diagnostic value.

All the above morbid appearances, including various kinds of new formations, such as cancer, sarcoma, etc., may affect the anus and rectum of men, who have never been given to pederasty.

In my opinion the same minimum diagnostic value is to be attached to a certain degree of prolapsus of the mucous membrane of the rectum, to which Tardieu draws attention, observing, that "the stretched mucous membrane of the lower portion of the rectum near to the orificium ani forms folds, and assumes the appearance of a slightly raised thick pad or roll. In other cases the folds of the mucous membrane resemble excrescences, which sometimes attain to such a development, that they form inspissations somewhat similar in appearance to the labia minora of the female genital organs, and open out when the orificium ani is stretched."

I have never noticed in Cynedes an appearance of this kind, which sometimes accompanies fresh cases of prolapsus recti, which have no connection whatever with a pederastic origin.

Another consequence of a considerable relaxation of the sphincter is noticeable in a certain incontinentia alvi, particularly of fluid fecal matter and gases. This involuntary evacuation causes a continual soiling of linen which in the case of boys may sometimes be considered as the first indication upon which experienced schoolmasters may recognize vicious habits in their pupils.

Further, the skin in the neighbourhood of the anus is moistened and often irritated and even inflamed by the contact of this excretion, which keeps up a continual state of humidity. All this tends to maintain such a state of uncleanness and so disagreeable and disgusting a smell in the affected parts, that it is not possible to imagine that their aspect can produce on the beholder any other feeling than that of disgust and repulsion.

The most unmistakeable, but at the same time the rarest symptom of sodomy is the appearance of a primary syphilitic induration (Ulcus induratum) in the neighbourhood of the anus, or in the rectum, it having been proved, that the first appearance of syphilis always shows itself at the very place where the infection has originated.

I have most frequently observed the appearance of an indurated ulcus in Cynedes at the point of transition of the skin into the mucous membrane, particularly on the anterior wall of the intestine, less often in the neighbourhood of the radial folds. The presence of an indurated chancre further away from the orificium ani, for instance upon the nates, the perineum, the posterior surface of the scrotum, can no longer serve as an undoubted proof of sodomy.

Considering the absence of pain in an indurated chancre of the anus, and the small amount of excretion proceeding therefrom, this primary appearance of syphilis may often escape the notice of the patient, or be taken for a simple fissure or laceration, etc.

It is only when secondary or subsequent accidents appear, when the disease, left without proper treatment, degenerates into syphilitic ulcerations, affections of the mucous membrane, of the scalp, etc., that the patient at last applies for medical assistance.

But at this period it is far more difficult to determine, in what way the infection has taken place. The indurated ulcus is by this time generally healed, leaving a hardly perceptible scar, or its place is taken by an appearance of secondary syphilis, which has developed itself at the same point (transformatio in situ), for instance by a wetting papula which has broken out over the entire anal region. With such secondary syphilitic appearances it is extremely difficult to determine exactly where the disease first showed itself, that is to say the starting-point of the primary syphilitic induration.

The discovery of the way and manner of the syphilitic infection then remains uncertain, as the appearance of secondary syphilitic symptoms in the neighbourhood of the anus does not in the least prove that the infection commenced there; it is known that secondary symptoms in the neighbourhood of the anus may appear after different modes of infection, for instance by the genital organs in normal coitus, or by the mouth in kissing, and so forth.

No doubt that during the further development of syphilis the disease very often manifests itself in various affections in the neighbourhood of the anus and of the rectum, in the form of secondary and hereditary ulcers, contraction of the rectum, etc. Nevertheless, here also, the localization of consecutive syphilitic appearances in the anus and rectum cannot be taken as proof that the infection originated in these parts.

Therefore it is only the positive presence of an unmistakeable primary syphilitic induration, or an ulcus induratum in the rectum, or in the immediate neighbourhood of the anus, that can be taken as a patent proof of sodomy.

It may then be asked:—but is it not really possible to be syphilitically infected through the anus otherwise than by way of sodomy?

Most certainly such a possibility cannot be theoretically denied; but, limiting myself solely to facts, I must say, that all the indurated chancres of the rectum I have hitherto observed—and this is no small number—had sprung from sodomy.

I know of one case only, in which a chancre showed itself at the spot of an operated fistula, caused either by the want of due care on the part of the surgeon, who had operated with unclean instruments, or by the fault of the assistant who had changed the dressing.

The grown-up pederastic patient, particularly the mercenary Cynede, usually seeks to explain his contamination, by saying that he had gone to the closet after a person affected with syphilis who had just quitted it, or had seated himself undressed in a bath-room, which had been just quitted by a sick person, or had accidentally put on linen belonging to a syphilitic.

Such and other inventions make it easier to the patient to say what ails him, but they are not in the least plausible, for it is neither in the watercloset, nor in the bath-room, nor by wearing an affected person's linen that the syphilitic virus can penetrate through the sphincter, thereby attacking the mucous membrane of the rectum, and that the spot affected can be inoculated with syphilitic virus.

It generally happens that after a certain lapse of time, when the confidence of the patient has been gained, he makes a further admission, which explains the real manner of the infection.

But if a primary syphilitic induration in the anus and in the rectum constitutes a sure evidence of sodomy, the same does not at all follow for the presence of a chancroid or ulcus molle in the same place.

On the contrary even, the formation of a chancroid in the anus by means of sodomy is of extremely rare occurrence, whereas the formation of such ulcerations around the anus from other causes may very frequently occur.

This depends partly upon the fact, that the chancroid does not usually appear alone, but more generally accompanied by several others, which cause pain and suppurate a great deal. On the one hand the pain and the multiplicity of the ulcerations make it impossible for the patient not to be cognizant of his condition, and on the other they render the sodomitic act extremely difficult. There are no doubt exceptions.

In the beginning of the last scholastic session I had under clinical treatment at the Imperial Academy of Medicine a young Cynede of 14 years of age, who was afflicted with an enormous phagedenic ulcus molle recti.

The ulceration occupied the entire neighbourhood of the anus, and penetrated to the depth of 4 centimetres into the rectum, causing the patient fearful pain at each act of defecation. The disease had its origin in a fissure at the point where the skin changes into the mucous membrane of the rectum and had gradually invaded the entire region around the anus, without being accompanied by the formation of any soft chancres on the genital or other parts of the body.

But I must repeat—that such observations are exceptions. The ulcera mollia appear the most frequently at the anus simultaneously with similar ulcers on the genital parts and originate by auto-inoculation in the radial folds. Whilst the suppuration from a soft chancre comes upon parts of the surface of the patient's body where the epidermis happens to be abraded or lacerated, it there easily inoculates itself, and causes the formation of a fresh chancre. It is precisely this auto-inoculation of the pus of an ulcus molle that serves to explain the multiplicity of such ulcers on the body of a patient, as well as their appearance on the anus after having originated by first infection on the genital organs.

That is why, on women, who in coitu have contracted ulcera mollia on the exterior genital parts, when the suppuration from the same flowing down through the short perineum attains the radial folds, should it there meet with any accidental laceration or fissure, it immediately inoculates the same. It has been noticed that, among the patients in the Kalinkin hospital, particularly among those who are not subject to the supervision of the medical police committee, i.e. who are not public prostitutes and therefore often previous to their admission remain for a long time without proper treatment, the ulcera mollia near to the anus are of daily observance; in these cases sodomy of course is altogether out of the question.

It may be here remarked that in St. Petersburg, particularly among the women, as also among prostitutes, sodomy is of extremely rare occurrence.

On men the overflow of the suppuration from the ulcers on the penis is generally stopped on its way down towards the anus by the scrotum, and therefore soft chancres near the anus are far less frequently observed on men, than on women. Nevertheless, scratching the anal region with the finger which may have retained upon it venereal matter after dressing an ulcer, may sometimes be the cause of an auto-inoculated chancre in the radial folds.

It may happen that the original ulcers on the genital parts are healed, whereas the soft chancres at the anus, where they are unfavourably situated with regard to cleanliness and facility of dressing, still subsist. Therefore, when ulcera mollia are discovered in the radial folds, care must be taken to examine closely so as to ascertain whether their origin may not be traced to previous chancres on the genital parts.

Accordingly an ulcus molle can only be exceptionally taken as a proof of sodomy, when it has originated first in the rectum, without similar chancres having previously existed on the genital parts, or so existing simultaneously.

A still greater rarity is sodomic gonorrhœa of the rectum.

I have only twice altogether had occasion to observe genuine, acute gonorrhœa of the rectum and in both cases on young prostitute Cynedes (bath-servants of from 15 to 17 years old). No doubt the appearance of acute gonorrhœa in the rectum of a young individual is almost a convincing proof; only it is needful to know how to distinguish gonorrhœa from the humid exudation and from the catarrh caused by worms, which in fact is not very difficult.

What is far more difficult is to differentiate blennorrhœa from the traumatic irritation of the rectum, which is observed on onanists who introduce divers objects into their anus, commencing with pencils and finally coming to glasses and bottles. [1] In grown-up subjects, particularly in old men, suffering from inveterate hemorrhoids, fistulas, prolapsus recti and catarrh of the rectum, with a copious purulent discharge, resembling that of gonorrhœa, may often be observed. Therefore it can only be on young subjects, who present no symptoms of other affections of the rectum, that a purulent catarrh of the rectum, in the acute form, may be taken as a proof of sodomy.

Older observers, Zacchias, for instance, reckoned as certain signs of sodomy the formation of verrucose excrescences, or so-called pointed condylomas, or more properly

---

[1] Moraud, Collection de plusieurs observations singulières sur des corps étrangers, les uns appliqués aux parties naturelles, d'autres insinués dans la vessie et d'autres dans le fondement. Mém. de l'Académie royale de chirurgie, 1757, T. III, p. 620.

speaking papillomas, in the neighbourhood of the anus.

Even the Roman satirists allude to "Crista" and "Mariscae" as evidences of sodomy. Thus Juvenal says in one of his most celebrated poems:

> "....Sed podice laevi
> Caeduntur tumidae, medico ridente, mariscae."

(But from your smooth behind are cut the swollen piles, the surgeon grinning the while).

Quite true I have had occasion to observe in habitual Cynedes, particularly prostitutes, who sometimes accomplish the sodomitic act several times in one day, the formation of verrucose excrescences on the border of the orificium ani, on the radial folds of the skin and even on the walls of the rectum. But these formations are in fact of yet more frequent occurrence quite independently of sodomy. Such papillomas may be observed in children who have catarrhal inflammation of the rectum resulting from the presence of worms, in adults affected with hemorrhoids and in elderly men subject to prurigo of the anus.

At times it is quite impossible to discover the cause of the continual growth of such papillomas in the neighbourhood of the anus.

Consequently the presence of verrucose excrescences at the anus cannot by themselves be considered as in any way a sign of sodomy, and it is only when they are met with in combination with other characteristic deformations, that they may sometimes serve partly to confirm the existence of sodomitic habits.

It remains now for me to mention one more peculiar symptom, first described by Casper, and particularly insisted upon by him; that is a peculiar conoidal sinking of the nates towards the anus. "A posterior of this kind does not present the usual hemispheres, but the inner side is flattened to a distance of from $1^{1}/_{2}$ to 2 inches from the anus, and there results a certain hollow between the nates, a conoidal cavity. This cavity is almost constantly to be found in habitual passive pederasts."

In my opinion the complete or incomplete juxtaposition of the nates, as well as the greater or less convexity of their inner surfaces, depends first of all on the stretched condition of the nates, on the age of the subject under exploration, on the position he occupies whilst undergoing examination, on the tension or state of relaxation of the glutæi muscles, and least of all on pederasty.

Thus, according to the different postures the subject under examination is made to take, will a conoidal hollow be observed at one moment and at another will have disappeared.

Even the etiology of this characteristic is incorrect. According to Casper the inner surface of the nates is supposed to become flattened, in consequence of the tension produced by the intromission of the sexual member into the rectum.

But in reality nothing of the sort takes place, because, in order to facilitate perfect intromission into the rectum, the nates are always held asunder with the hands, and are therefore subject to no pressure from the penis.

Casper pretends that this characteristic indicated by him is one of the most conclusive of all the uncertain symptoms of passive pederasty. It seems to me more conformable to truth to paraphrase Casper's assertion, and to say, that the conoidal hollow between the nates is the least conclusive of all the other certain signs of pederasty.

After having passed in review the signs of passive pederasty, we will now draw attention to those appearances which are discovered by exploration in cases of quite recently committed sodomy.

All observations of this nature belong mostly to the category of sodomitic rape, committed either by an impulsive pederast during an access, or by an aged man in a fit of senile dementia, or else by a congenital active pederast presenting in a high degree psychical degeneration; for it is only when the intelligence has fallen very low, or when the paedicator is under the influence of well-defined psychical disturbance, that the sodomitic act is suddenly committed with any degree of violence. On the contrary, in the immense majority of cases, as already observed, the Cynede is gradually taught his part. Sometimes months are first past in preliminary manipulations, and it not unfrequently happens that at the time when the complete intromission is accomplished for the first time, the Cynede already possesses the distinctive signs of sodomy: the infundibuliform widening out of the anus, relaxation of the sphincter and other characteristic signs.

We will therefore here speak of the signs of freshly committed sodomy on an unprepared subject, in other terms, the signs of so-called sodomitic rape, or more correctly sodomy with applied violence, as the real rape, that is to say the accomplishment of the entire act from beginning to end, against the will of the Cynede is impossible.

There are certainly cases, where little children of from two to three years old, or more, have been sodomised by force, either with the assistance of several persons, or else, while in a state of insensibility; but these are rare exceptions.

In most cases the commencement of the act is submitted to with good grace, and it is only occasionally that it ends with the employment of a certain amount of force.

The injured subject, generally a boy or youth, complains during the first few days following the act of a feeling of soreness when in the act of defecation and of pain at the anus when sitting on a hard seat; the gait is somewhat changed, the legs are more separated than usual; later on the pain gradually diminishes and gives place to itching. The exploration of the anus shows an inflamed redness of the surrounding skin, swelling and numerous slight lacerations in the mucous membrane of the rectum, with slighter ones on its border, on the exterior skin together with patches of ecchymosis. There is often oozing from the anal orifice a purulent-like fluid mixed with blood. The introduction of the finger into the rectum is very painful, causing contraction of the levator and sphincter muscles. It is only the employment of considerable force with a large sexual member that can in these first few days cause relaxation of the sphincter by either inordinately expanding or actually tearing it.

Then comes another symptom,—Incontinence of alvine matter and involuntary escape of gases.

Finally, when the highest degree of force has been used, under the influence of a violent trauma, the cellular tissue of the epidermis round the anus may become inflamed, which may give rise to the formation of abscesses and of consequent fistulas.

Besides these various appearances at the anus and rectum, when there has been a considerable amount of violence employed, certain deformations may be observed in the party concerned, particularly in the genital parts. It is even possible, according to the extent of these deformations, to form an opinion of the more or less degree of force that has been employed in the act.

Not unfrequently there may be observed œdema of the prepuce, fissures on its border, rupture of the frenulum, lacerations and extravasated blood on the scrotum and perineum.

In a case recorded by Tardieu the entire skin of the penis, from the very root, was torn and turned inside out like a glove. [1]

In another case, accompanied by murder of the victim, the scrotum was found to be much swollen and there was considerable hemorrhage.

In a third, in which two men had committed sodomitic rape upon a child of 3 years of age, afterwards murdering it in the most barbarous manner, there were found upon the body of the unfortunate victim traces of deep teeth marks and lacerations made with finger-nails on the skin at the root of the member and on the scrotum.

Casper [2] also records a case of sodomitic rape on a little boy of five years of age accompanied by an attempt to strangle the victim, in which the child's prepuce was rent, so that the cellular tissue was visible as far as the corpora cavernosa of the penis.

[1] Tardieu, *loc. cit.*, p. 267.
[2] Mierzejewski, *loc. cit.*, p. 228.

But far more serious injuries to the genital organs are noted in the case described by Dr. Marquisi.[1]

But with every succeeding day the above-described signs fade, disappear, lose their signification, and it becomes almost impossible any longer to recognize them two or three months after the criminal assault—except in exceptional cases, for instance when there has been syphilitic infection.

I have reason founded on fact to believe that a sodomitic act committed once or twice with a certain amount of violence on a boy of from 10 to 11 years of age would after two or three months leave no visible traces of deformation on the anus and rectum. I should therefore not place the same confidence, as Dr. Espallac[2] did, in the tale of a young girl, 12 years old, that she had been subjected only twice to the sodomitic act, and should indeed rather be inclined to doubt it, the more so that two months later on the least separation of the nates there was relaxation of the sphincter and gaping of the anal orifice.

The series of facts we have brought forward

---

[1] Giraldès et P. Horteloup, Sur un cas de meurtre avec viol. sodomique. Ann. d'Hyg. et de Méd. lég. 2e Série, T. XLI, p. 419.

[2] Tardieu, *loc. cit.*, p. 227.

proves to a certainty that habitual passive pederasty, and still more easily when of recent or accidental occurrence, can be recognized on exploration by indubitable objective signs.

Another question is, whether the above-described deformations of the anus, rectum and genital organs, perceptible on exploration, are to be accepted as sufficiently conclusive to recognize all cases of passive pederasty? We shall examine this question further on, and will now see whether there exist signs of active pederasty.

Of all observers Tardieu is the only one who answers this question in the affirmative. According to his opinion, founded on 133 observations, the virile member of active pederasts presents certain deviations from the normal type, a conclusion which may in certain cases be admitted.

Tardieu finds, that habitual active pederasts have mostly a very slender, poorly developed penis with a small gland, and that it gradually diminishes in size from the root towards the gland, giving it some resemblance to a dog's penis. In rare cases it is on the contrary unusually big, and then it is the gland alone that takes a tapering form, whilst the stem is as it were twisted on its axis, so that for instance the orifice of the urethra is on one side or presents a lateral slit.

Tardieu, in order to explain the above-described deformation, particularly the tapering of the gland and the twisting of the stem, attributes the same to the pressure of the sphincter muscle and to the repeated considerable forcible effort and screw motion required to obtain the intromission of a largely developed member into the rectum.

The insufficiency of Tardieu's explanation is evident.

The principal sign, consisting in the peculiarly small and slender dimensions of the entire member and of the gland, is not at all the result of habitual sodomy.

But it is not the explanation that is of importance, but the fact.

It appears to me that Tardieu's observation is to some extent correct, but that he concluded too hastily upon isolated cases and that his explanation is therefore altogether wrong.

I am able to maintain from personal observation that the majority of freely acting paedicators show no perceptible deformations whatever on their genital organ, which could enable one to divine their vicious propensity.

All acquired pederasts, active pederastic prostitutes, morbidly demoralized old men, in one word the majority of active pederasts, exhibit no difference whatever of their sexual parts from those of normal beings.

There may be, however, some relatively rare cases of congenital active pederasts who present visible deviations in the development of the sexual organs.

I have already remarked, that this form of sexual pederasty is in most cases caused by a more considerable degree of degeneration than is to be found in the more commonly found congenital passive pederasts; that is why in such cases a close inspection will discover various evidences of interrupted development, among others also defective, or irregular formation of the sexual parts.

In this sense Tardieu's observation is correct.

However, taking into consideration the general fact that the study of the signs of physical degeneration is as yet in its infancy, so none of them ought to be neglected, particularly one so clearly defined as defective or irregular development of the genital organ.

For instance, out of four active pederasts known to me at the present moment, there are two who have a remarkably thin penis with a tapering gland; the third, in consequence of an unequal asymmetrical development of the corpora cavernosa, has the member slightly inclined on one side; and in consequence of a similar defective formation of the corpora cavernosa of the urethra he has the gland so turned that the orifice is not from above below, but oblique. The fourth pederast suffers from congenital phimosis and atrophy of one of his testicles, whereas the aspect and size of the penis are apparently normal. On all four, besides the above-mentioned deviations, there are observable other clearly marked signs of physical and psychopathic degeneration.

No doubt it is altogether impossible to stamp as active pederast an individual because he happens to show the above-described deformations of the sexual parts.

Exactly the same appearances may be observed on persons with normal sexual functions. But when it is a notorious pederast who exhibits the above-mentioned abnormalities, we have then almost invariably before us, not a case of acquired, but of congenital perversion.

It is therefore evident that the ocular

examination of the sexual parts, and that of the body in general, furnishes no true sign upon which a single proof of active pederasty might be founded with even approximative certainty. Nevertheless all deviations from the regular formation of the sexual parts in pederasts would seem to point to congenital perversion of the genesic instinct.

It is not possible to recognize pederasty in general by means of the signs correctly given by Tardieu; but by their means the etiology of abnormal sexuality in pederasts can be determined with certainty.

We have now passed in review all the known signs of pederasty. Do they suffice, if not in all, at any rate in the majority of cases, to determine the sexual perversion in question?

Unfortunately, such is not the case.

Putting the active pederasts out of the question, who, as we have just seen, are not recognizable by any distinct marks to be found by exploration, it is equally difficult to diagnose all passive pederasts as such.

Pederasty, in the general sense of the word, does not, as we have previously observed, always manifest itself in sodomy. Those pederasts, who do not actually accomplish the sodomitic act, but discharge their semen *inter fecces*, like the Fellators, will exhibit from that cause no deformations of the anus or of the rectum, nor of the genital organs. Passive pederasty also, when it is beginning to be resorted to by some demented old man, even when he is afflicted with hemorrhoids, will leave no characteristic signs upon the anus. These signs will also be wanting on the periodical pederast, who accomplishes the act only after long intervals during which the deformed parts have plenty of time to resume their normal condition.

But, in the limited number of cases in which the existing physical signs point to sodomy, are they sufficient to decide the principal question that interests us—viz. whether it is a congenital taint or a vicious habit?

Certainly not!

Those signs, which have been hitherto observed on pederasts, cannot enlighten us as to whether the individual under examination is a demoralized mercenary Cynede who earns money in this shameful manner, or a subject under the influence of a serious malady, the morbid symptoms of which are manifested in sexual perversion; or again if it is an unfortunate being, whose development is behindhand, and who, from his birth, has been deprived of the faculty of regular sexual connection.—To sum up, the deformations of the anus, of the rectum and of the genital parts taken alone leave the principal question undecided—as to what kind of pederasty we have to do with.

The resolution of this important question demands an entirely different method of inquiry.

When it is the case of a youth we must first of all endeavour to find out, on the basis of heredity, of physical degeneration, of psychical aberration, etc., whether he is a healthy or a psychopathic subject.

When in the first case there exist indications of pederasty, then he is an acquired pederast; in the latter it is most probable that the vice was born with him.

The acquired perversion of the sexual instinct is proved, when the youth assumes alternately the active and the passive part, and that he can at the same time have normal connection with women.

The congenital form, on the contrary, is manifested in the great majority of cases, particularly in youths, in that they are exclusively inclined to passive pederasty, and are not capable of accomplishing connection with women.

The rare cases of congenital active pederasty, which are always accompanied by well-defined psychical aberrations, sometimes show also an abnormal formation of the genital parts.

Subjects of this kind are not only in general deprived of the power of having sexual intercourse with women, but they feel a positive hatred against them.

When it is a grown-up man who is subjected to examination, the question becomes far more complicated.

Here, it is necessary to collect facts concerning heredity, and initial anamnesis, to follow up stage by stage the life that has been led, more especially during the period of puberty, combined with a full enquiry into all the physical and psychical peculiarities of the subject under examination, so as to endeavour to decide the question whether we have before us a case of congenital or of acquired sexual perversion.

In the first case, we can ascertain, from a knowledge of facts revealing the time and conditions of the beginning of the vice, the frequency and uniformity of the repetition of the act, etc., if the subject under examination belongs or not to the more dangerous but less punishable class of periodical pederasts.

When the presence of congenital sexual perversion is contradicted by established facts, it is necessary to discover if the appearance noted is a symptom of incipient progressive paralysis, or an accessory phenomenon, a so-called psychical equivalent of epilepsy, or lastly if it is a symptom of premature senile dementia.

It is only after excluding all the above-named morbid conditions on positive grounds, based on facts deduced from careful enquiry and observation, that we can be permitted to affirm with probability the depravity, moral corruption and absolute, voluntary and premeditated licentiousness of the subject under examination.

The decision of the above questions meets with the greatest difficulty when we have to do with elderly men, and here the most difficult task consists in the distinction of pederasty as a symptom of incipient senile dementia from positive vice, which is for ever seeking for new means to revive the expiring sexual power.

The gradual diminution of the mental powers and the deadening of the feelings only attract the attention of friends or of strangers when they are approaching the final stage.

The mental debility passes almost imperceptibly from the usual so-called weak-mindedness into complete ruin of the faculties of the mind.

A thorough imbecile is unable to grasp the meaning of a syllogism in its entirety. Each premiss is to him a separate idea, and it is beyond the power of his comprehension to connect it with the preceding one or with the one which succeeds, to retain them in his memory and from them to deduce a conclusion.

In this clearly defined form imbecility is easily diagnosed. But some degrees higher, we come upon a subject, who is able to reason, to draw conclusions, to connect together several ranges of thought, but who does not possess the faculty of realizing the conception of a thing in its entirety, to abandon a course of ideas once adopted and to examine a subject from different points of view.

From such a mental debility there is but

one step to absolute obliteration of the intelligence.

The subject may be treated logically and generally, but with continual wanderings from the fundamental idea, and becomes so confused by the introduction of details and accessories, as to lose all sequence, leading therefore to no settled result.

An empty, apparently logically connected, sometimes witty gossipping habit, forms one of the most frequent symptoms of an incipient intellectual decadence due to age, or in subjects who are exhausted by all sorts of excesses. It very often happens that the same symptom combines with it the degree of mental poverty, that is known in society under the name of "average intelligence."

The absence of full reasoning power, insufficient critical acumen, inability to distinguish the important from the unimportant, want of independent effort to discover the cause and manner of things, intellectual one-sidedness, loss of creative power and of originality of thought—all these are so many tangible symptoms of different degrees of mental impoverishment and decay of the understanding.

It is necessary to have known the subject for a long time, in order to establish a sure basis of comparison between his previous normal intellectual activity, which does not solely consist of thinking and creative processes, and his actual morbidly depressed mental powers.

Receptivity towards exterior impressions and their implantation in the consciousness, memory, faculty of comprehension, logical processes, sentiment—all these have to be taken into consideration before coming to a conclusion.

It is still more difficult to recognize a deadening of the senses, for here we possess no sure exterior symptom of loss of sensibility; and this is no less true with regard to hyperæsthesia, which does not show the strength of the sensibility, but the manner of its expression.

It is comprehensible, that with the lowering of the intelligence and the deadening of the sensibility the two most powerful means fail for resistance against the passions, wherein the degree of virtue is recognized by the understanding and estimated by the sensibility.

To these two factors, which constitute the dominant note of the general character of the feeble-minded, there comes further in addition increased sensuality, morbidly exaggerated erethism, under whose influence the subject sometimes becomes guilty of a whole series of criminal acts.

In fact, as soon as it is proved that the individual in question had previously rejoiced in the exercise of normal sexual functions, that the same gradually diminishing, finally disappeared, but after certain intervals revived with renewed vigour but in a perverted form, there can be no doubt that we have then before us a case of commencing senile dementia.

Above all, sexual desire not in keeping with his age is the very first symptom of a developing morbid condition of the subject.

The depraved individual seeks by changing the manner and means of accomplishing the sexual act, as well as in preparation for the same, a new way to increase and prolong the erethism.

On the contrary the old man morbidly affected seeks exclusively to find in a deviation from the normal function a better and more complete satisfaction of the lust that is continually tormenting him.

In accordance with these different objects the manifestation of the sexual instinct is usually somewhat different in the two cases.

The depraved individual employs every

means in his power that can contribute to increase his lust.

Sight, touch, hearing, smell, even sometimes taste, all the senses one after another, or all together, are excited to a certain extent, in order to raise the erethism to its highest degree of intensity.

It is under these excitations that passive pederasty shows itself, as a casual accessory phenomenon, as a new excitant, which may serve to heighten the erethism, which then finds its satisfaction in normal connection with a woman.

Sometimes the use of exterior and interior stimulants is added, and also the reading of pornographic books, and so forth.

Notwithstanding his apparent unrestrainable lust, the depraved individual is capable on occasion of recovering mastery over himself, he can show himself as a model husband, or the austere magistrate, condemning vice.

In genuine cases of senile dementia, the patient on the contrary does not seek for excitement. Unable always to dissimulate his sensual desire, he leads the conversation to the subject of sexual connection, becomes cynical, and even allows himself to make indecent gestures. All this is not done for the purpose of causing excitation, but from the desire to satisfy the continually increasing sexual erethism that torments him. For the same reason he repudiates normal connection with women, which cannot satisfy his morbid lust.

He either changes it, in turning from women to little girls and children, or not finding there sufficient satisfaction, becomes an active or a passive pederast.

And now, in all the various means he resorts to in order to satisfy his morbid sensual, craving lust, there is gradually developed an excited desire to make the victim suffer physical pain.

In cries, groans and convulsive movements the patient sees something stronger than the voluptuous spasm, and in this way he endeavours to find satisfaction.

Under the influence of this feeling, from biting and scratching with his nails, he comes to cutting and decapitating, impelled by one desire only, that of satisfying by any means his ever increasing morbid licentiousness.

The preceding facts show, how the manner and mode of manifestation of the genesic energy of a simply depraved individual differs from the insatiable and never ceasing lust of the weak-minded old man, in his desperate efforts to satisfy the morbid irritation of his sexual sense.

In the latter, in most cases, the sexual desire will appear simultaneously with feelings of an impeding character, which in a healthy subject would suffice to damp his desire, such as, for instance, wickedness, anger or the wish to subject his victim to pain, to witness its sufferings, to hear its cries, to feel its death throes.

The old man morbidly affected will sometimes kiss his victim, at the same moment tearing its body open, from which the blood streams out, works himself into a passion, uttering fearful threats and at the same time accomplishing the act of coition.

The depraved individual, on the contrary, attaches the greatest importance to exterior circumstances, seeks to keep away and avert all unpleasant feelings, being so to say occupied in concentrating himself in voluptuousness, is attentive to small things, capricious and full of pretention. Whereas the weak-minded old man performs the sexual act under the most various conditions, enjoying it according to his taste, sometimes in the most unworthy and disgusting manner.

The depraved subject, on the contrary, cleans, washes and perfumes his victim with pleasant aromas, combs and dresses him according to his taste, whereas the sick old man will commit a rape on a ragged dirty street urchin in a stinking stable.

But, however different the forms of the manifestation of senile dementia may be, it must be considered that there are numerous indistinctly marked transition stages, in which it is impossible to determine where vice ceases and disease begins.

This is the more difficult that very often those who have led a loose life in youth become the victims of senile dementia later on.

It is then that vicious habits gradually and imperceptibly change into morbid symptoms.

I believe in general, and with great probability, that the beginning of the malady is indicated, when to increased sexual desire there is added instincts of cruelty. the wish to subject the victim in the acme of the voluptuous spasm to physical pain, to hurt and wound him.

Of course it is only the continuance of cruelty during the sexual act that is characteristic, and not casual bitings, blows and wounds, which may be inflicted by a sanguine subject in the heat of erethism, or when in a state of inebriety.

I have here but briefly indicated those circumstances, which may with more or less probability decide the question, in a given case of pederasty, whether we have to do with a congenital fault of development, or an acquired vicious habit, the expression of a deep-rooted depravity, or a nervous disease. This cursory review serves to show how completely such a subject must be examined, how carefully and how long he must be kept under observation, so as to follow him in the minutest details of his whole life, to see what may have been the influence of education, of example, of the maladies he may have had to overcome, etc. To this must be further added the slightest details concerning the development and activity of his sexual instincts, the mental and intellectual sphere in which he lives, his social and family surroundings.

To the above it is necessary to join the closest knowledge of the state of health, character and habits of the parents of the subject under observation, and in fact of all his blood-relations,—in one word, the most complete data possible should be obtained on heredity, surroundings and other possible factors of degeneration—and it is only then that we can with more or less certainty determine what is the type of sexual perversion we have before us, and find out the cause of the aberration, as far as the actual state of science will permit us.

If every medico-legal case of pederasty is treated in this manner, it is easy to understand, what a quantity of details must be collected for the enquiry, and what a difficult and complicated task the medico-judicial expert will have to accomplish.

I therefore think, if the facts I have just produced have sufficient convincing force, that in future pederasty is no longer to be solely attributed to insatiable licentiousness and depravity, nor the examination to be limited to that of the anus and genital parts of the subject.

I know very well that nothing is more hurtful to the mentally afflicted who commit crimes while in a state of aberration, than false philanthropy or mercenary eloquence, which seeks to generalize the principle of irresponsibility, and to make invented imaginary mental aberration stand for real undoubted guilt. It is far from my intention, in pursuing this study, to supply vice with a weapon wherewith to combat the law.

Hitherto, in examining deviations of the sexual instinct, too little attention has been paid to congenital taint and disease.

My object was—as far as possible—to insist upon the addition of these factors to the subject under enquiry, and therefore I have said as little as possible about moral depravity, although far from undervaluing the importance of vice as a factor, wishing merely to differentiate it from the pathological complications which make it so difficult to establish an impartial estimate of the degree of moral decline of a really depraved subject.

The mingling of disease with vice always diminishes the importance of the latter.

For instance, can the gradual leading astray of a child, who is taught little by little to give itself up to vice, be compared with the rough and brutal commission of a pederastic rape?

In the first case there is neither threat nor violence; everything goes on slowly, and develops gradually, coming to the end as it were of its own accord. The vice is usually surrounded with such precautions, that it seldom comes before the judicial bench, and still more seldom meets with punishment.

In the latter case there is not only an immoral act, but the sense of shame of the victim has been outraged in the highest degree, his will overpowered and his body injured. When in both cases the same impelling motive is taken for granted, there can be no doubt that an error in the bringing up of the subject will appear as a slight offence compared with violent rape. But in fact the bringing up of the victim purposely for the vice is the greatest and most abominable expression of premeditated depravity, whereas pederastic rape is more often a manifestation of senile dementia.

In the presence of a symptom of disease, the most unbridled demoralization fades and its liability to punishment is correspondingly diminished.

The more precisely and clearly the field of the vice in question is circumscribed, the more evident do the slightest offences against decency become, offences which might otherwise disappear in the mass of pathological facts, and their liability to punishment becomes more clear.

On the other hand the discovery that the

deviations and perversions of the sexual instinct may be the symptom of a psychical disturbance is of great importance.

It is of course beyond the actual state of our science, to distinguish between all possible cases of vice and of disease; we are similarly just as little able always to distinguish crime from psychosis.

Maudsley says that "between crime and insanity there is a border-land, where there is on the one hand a small dose of insanity and a large percentage of crime, and on the other hand a small admixture of crime and a large proportion of insanity."

This is perfectly applicable to the connection between vice and malady. There undoubtedly lies between the two a place of transition, a border-land of "morbid depravation", where it is difficult, if not altogether impossible, to determine the proportion of voluntary, premeditated vice and that of hereditary predisposition, or the manifestation of an inherited morbid condition. But leaving aside the cloudy, undecided exceptional cases, we have nevertheless before us a scientific basis whereon to establish a distinction between disease and vice,—a basis, which on more complete development in this direction will, in my opinion, furnish more useful practical results than the numerous treatises founded on principles that are written in defence of morality and sobriety.

When once a malady is clearly recognized, hardly anyone will wish to imitate its symptoms, particularly if it is generally known and placed beyond doubt, that certain manifestations of the same always indicate an abnormal condition of the nervous centres and a weakening of the intellect.

At present vice not only appears seductive to most people because of the force, novelty or diversity of the feelings excited, but it also gives to the libertine in the sphere of sexual activity a certain tone of epicureanism, of originality, of vicious repute and of superiority over other beings, who appear to be less perfectly developed in contrast with him, but more moral and more abstemious.

In society the idea prevails that the taste which has become satiated with every ordinary enjoyment inclines to refined licentiousness, seeking sexual satisfaction for more developed and more supreme enjoyment.

Such an apparent inventive genius in vice inevitably presupposes something superior and more complete than the usual way of satisfying the sexual instinct. Therefore the consciousness of vicious propensities and of sexual depravity raises certain ignorant, or weak-minded individuals in their own estimation and in that of their circle.

This attractive feature of vice, which favours the imitation of morbid depravation, must lose its charm with the knowledge that vice in its most violent manifestations is the symptom of a morbid state, with a certain deadening, and not a refining of the feelings, and with an imperfect equilibrium of the nervous system which, far from leading to perfection, favours the development of mental disturbance and weakness of intellect.

In this respect the Law Courts may do Society a service of vast importance by disseminating abroad a body of sound opinion.

I am in complete agreement with Michelet when he says "that Jurisprudence must become a medical science, based on physiological facts, in order to determine the influence of inconscient, fatal impulses on voluntary acts."

I am convinced, that it is only the combined work of the Physician and of the Jurist—of the Investigator and of the Philosopher—which can at the same time discover the causes of these impulses and of their manifestations in acts, determine the limit between physiology and pathology in life, and furnish a solid basis for the improvement of the sane, the education of the morbidly inclined, and the cure of those afflicted with disease.

BIBLIOGRAPHY AND INDEX.

BIBLIOGRAPHY.

Albert, Dr., Friedreich's Blätter. 1859. III. p. 77.
Anjel, Ueber eigenthümliche Anfälle perverser Sexualerregung (Arch. f. Psych. Bd. XV. Heft 2).
Béraud, Les filles publiques de Paris et la police qui les régit. 1839.
Bartholomew's, St. — Hospital; Obstinate Priapism. Med. Times and Gaz. 1852.
Birket, I., Case in which persistent Priapism, etc. The Lancet. 1867. Vol. I. p. 207.
Blumer Alder, A case of perverted sexual Instinct. American Journ. of Insanity. 1882. July.
Blumröder, Ueber Lust und Schmerz. Friedreich's Magazin für Seelenheilkunde. 1830. II.
Bordier, Étude anthropologique sur une série de crânes d'assassins. Revue d'Anthropologie. 1879.
Briere de Boismont, Gaz. médicale de Paris. 1849. 2 Juillet.
Broca, Masturbation invétérée, Infibulation. Bull. de la Soc. de Chir. de Paris. 1865. T. 5. S. 2. p. 10.
Brouardel, Étude critique sur la valeur des signes attribués à la Pédérastie. Annales d'Hygiène publ. 1880. No. 20. p. 182.
Bucher, E., Lehrbuch des gerichtsärztlichen Medicin. 1872. II. Aufl. p. 197.
Buffon, Histoire naturelle de l'homme. Puberté.

Canler, Mémoires de, Ancien chef du Service de Sûreté. Bruxelles. 1862.
Casper, Klinische Novellen zur gerichtlichen Medicin. Berlin. 1863.
— Praktisches Handbuch der gerichtlichen Medicin (russische Uebersetz.). St. Petersburg. 1872.
— Ueber Nothzucht und Päderastie. Vierteljahrsber. f. gerichtl. Medicin. 1852.
Charcot et **Magnan**, Inversion du sens génital. Arch. de Neurol. 1882.
Cohn, Prof. Hermann (Breslau), Augenkrankheiten bei Masturbanten. Neurologisches Centralblatt. 1882. p. 63.
Coutagne, Notes sur la Sodomie. Lyon médical. No. 35, 36 1880.
Crothers, F. D., Inebriate Automatism. The Journ. of nervous and mental disease. No. 2.
Démeaux, Priapisme spontané. Annales de la Chir. franç. et étrang. 1841. Vol. 3. p. 403.
Demme, Buch der Verbrechen.
Descuret, La médecine des passions. Paris. 1860.
Diez, E. A., Der Selbstmord. 1838.
Dufour, Pierre, Histoire de la prostitution chez tous les peuples du monde, etc. 1851—54.
— Mémoires curieux sur l'histoire des mœurs et de la prostitution en France aux XVII et XVIII siècles. 1854.
Érections génitales morbides chez l'homme. Gaz. des Hôp. 1852. p. 10.
Eulenberg, Vierteljahrsschr. f. gerichtl. Med. 1878. Bd. 28.
Fahner, System der gerichtl. Arzneikunde. Bd. III. p. 186.
Filippi, A., Manuele di aphrodisiologia civile criminale e venere forense. Pisa. 1878.
Frentzel, De Sodomia. Erfurt. 1723.
Friedreich, J. B., Handbuch der gerichtsärztlichen Praxis. 1843. Bd. I. p. 271.

Garnier, P., Onanisme seul et à deux, sous toutes ses formes. 1884.
Giraldès et **Horteloup**, Sur un cas de meurtre avec viol. sodomique. Annales d'Hygiène publique. 1874. p. 419.
Gock, Von, Beitrag zur Kenntniss der conträren Sexualempfindung. Arch. für Psych. Bd. V. p. 564. 1875.
Guyot, Yves, La Prostitution 1883.
Griesinger, Ueber einen wenig bekannten psychopathischen Zustand. Arch. f. Psych. I. p. 651. Berlin. 1868.
Gysbrechts, Observation de priapisme. Journal de Méd. de Bruxelles. 1848. Vol. 7. p. 223.
Hammond, Sexual Impotence in the male. New York. 1883.
Hartmann, Pædicatorem noxium esse. Frankfurt. 1776.
Hartung v. Hartungen, M.U., Ueber virile Schwäche, etc. Wien. 1884. p. 188.
Hatté, Sur le Satyriasis ou Saturiasme. Recueil périodique de médecine, chir. et pharmacie. 1755. T. 2. p. 109.
Hofmann, E., Päderastie. Real-Encyclopädie der gesammten Heilkunde. Bd. X. p. 294. 1882.
Jacob, Curiosités de l'histoire de France. Causes célèbres. Paris. 1859.
Jeannel de Bordeaux, De la prostitution publique et parallèle complet de la prostitution romaine et de la prostitution contemporaine. Paris. 1863. 2me Ed.
Kaan, Psychopathia sexualis. Leipzig. 1844.
Kirn, Dr. L., Ueber die klinische forensische Bedeutung des perversen Sexualtriebes. Allg. Zeitschr. f: Psych. Bd. XXXIX. p. 216.
Klose, Ueber Päderastie in gerichtlich-medicinischer Hinsicht in Ersch und Gruber's Allgem. Encyclopädie. 3. Sect. 9. Theil. Leipzig. 1837.
Kowalewski, P. J., 1) Die Primäre Verrücktheit. Für Aerzte und Juristen verfasst. 1881 (russisch).

2) Forensisch-psychiatrische Analysen. 2 Theile;
für Aerzte und Juristen. 1881 (russisch).
Krafft-Ebing, a) Ueber gewisse Anomalien des Geschlechtstriebes, etc. Arch. f. Psych. u. Nervenkr. 1877. VII. p. 291.
b) Lehrbuch f. Psychiatrie.
c) Zur Lehre von der conträren Sexualempfindung. Irrenfreund. 1884. No. 1.
Krauss, A., Die Psychologie des Verbrechens. Tübingen. 1884. p. 173.
Lasègue, Les Exhibitionistes. Union médicale. 1877. 1 Mai.
Legrand du Saule, 1) La Folie devant les tribunaux.
2) Les signes physiques des folies raisonnantes. Annales médico-psychologiques. 1876. Mai.
Lombroso, 1) Vetzeni e Agnoletti. Roma. 1874.
2) Amori anomali e precoci nei pazzi. Arch. di psich. etc. 1883. VI.
Martineau, L., Leçons sur les déformations vulvaires et anales. Paris. 1884.
Maschka, S., 1) Unzucht wider Natur in Handbuch der gerichtl. Med. 1882. Bd. III. p. 176.
2) Prager med. Viertelj. Bd. 89.
Meibomius, Johann Heinrich, De Flagrorum usu in re veneria. Londini. 1770.
Meier, E., Ueber Päderastie im Alterthum in Ersch und Gruber's Allg. Encyclopädie. 3. Sect. 9. Theil. Leipzig. 1837.
Mende, Handbuch der gerichtl. Medicin. Leipzig. 1826. Bd. IV. p. 506.
Menesclou (affaire), Rapport de Lasègue, Brouardel et Motet. Annales d'Hygiène publique. 1880. p. 439. Paris.
Menière, P., Études médicales sur les Poètes latins. Paris. 1858. v. Juvenal. p. 351 et Martial. 433.
Mierzejewski, W., Forensische Gynäkologie. Hand-

buch für Aerzte und Juristen. St. Petersburg. 1878 (russisch).
Michea, Union médicale. 1849.
Montagne, E., Histoire de la prostitution dans l'antiquité. 1869.
Moraud, Collection de plusieurs observations singulières sur des corps étrangers appliqués aux parties naturelles. Mém. de l'Acad. royale Chirur. 1757.
Moreau (Paul) de Tours, Des aberrations du sens génésique. Paris. 1880.
Negris, De la dynamie ou exaltation fonctionnelle au début de la paralysie générale. Paris. 1878.
Numantius, Numa (Karl Heinrich Ulrichs), Anthropologische Studien über die mammmännliche Geschlechtsliebe, "Incubus und Gladius Furens". Leipzig. 1869.
Parent-Duchatelet, La prostitution dans la ville de Paris. 1857. T. I. p. 214.
Pitaval, Causes célèbres. T. VIII. p. 511.
Priapism, a case of, — requiring incisions. London Med. Gaz. 1830. p. 92.
Rabow, Zur Casuistik der angeborenen conträren Sexualempfindung. Centralblatt f. Nervenheilk. u. Psych. 1883. p. 186.
Rabutaux, A., De la prostitution en Europe depuis l'antiquité jusqu'à la fin du XVIme siècle. 1865.
Raggi, Aberrazione del sentimento sessuale in un maniaco ginecomasta. La Salute. 1882. No. 11. p. 86.
Récamier, Gibert, Excitation habituelle des organes génitaux. Accès de Satyriasis. Tenesme vesical contenu. Gaz. des Hôp. 1829. T. II. No. 44. p. 174.
Rey, J. L., Des prostituées et de la prostitution en générale. 1847.
Riche de la Popelinière (Le), Tableau des mœurs du temps dans les différents âges de la vie, avec note de Charles Monselet. Paris. 1867.

Rosenbaum, Die Lustseuche im Alterthum. Halle. 1839.
Roubasd, Traité de l'impuissance et de la stérilité. 1876. 3me édition.
Rousseau, J. J., Les confessions. Partie I. Livre I.
Rul-Oger, Priapisme et pollutions nocturnes. Journ. de méd. de Bruxelles. 1848. Vol. VI. p. 19.
Sade (Marquis de), La nouvelle Justine ou les malheurs de la vertu. Hollande. 1797.
Shaw, J. C. and S. N. **Ferris**, Perverted sexual Instinct. The journal of nerv. and mental disease. 1883. No. 2.
Schminke, Ein Fall von conträrer Sexualempfindung. Arch. f. Psychol. Bd. III. 1872. p. 225.
Scholz, Bekenntnisse eines an perverser Geschlechtsrichtung Leidenden. Vierteljahrsschr. f. gerichtl. Med. 1873.
Schopenhauer, A., Die Welt als Wille und Vorstellung. 1859. Bd. II. p. 641.
Schuring, De coitu nefando seu Sodomia. Gynäkologie. Sect. II. cap. VII.
Servais, Zur Kenntniss von der conträren Sexualempfindung. Arch. f. Psych. 1876. Bd. VI. p. 484.
Sprengel, Geschichte der Medicin. 2. Aufl. p. 83.
Stark, Ueber conträre Sexualempfindung. Allgem. Zeitschr. f. Psychol. 1877. Bd. XXXIX. p. 209.
Sterz, Dr., Beiträge zur Lehre von der conträren Sexualempfindung. Jahrb. f. Psych. Bd. 3. Heft 3. p. 221.
Stoltenberg, Diss. in prædicatorem noxium et infestum rei publicæ civem. 1775.
Suetonius Tranquillus, De vita Cæsarum.
Tamassia, Sull'inversione dell'instinto sessuale. Rivista sperini di freniatria e di medicina legale. 1878. T. XV. p. 97.
Tardieu, A., Étude méd.-légale sur les attentats aux mœurs. 7me éd. 1878.

Taxil, Léo, La prostitution contemporaine. 1884.
Taylor, Medic. Jurisprudence. 1873. II. p. 473.
Toulmouche, Des attentats à la pudeur et du viol. Annales d'Hygiène publique. 1868.
Venot de Bordeaux, De la pseudosyphilis chez les prostituées. Bordeaux. 1859.
Vidal et **Legrand du Saule**, Ann. médico-psychol. Vme Série. 1876. T. XV. p. 446.
Walter, Prof., Ueber geschlechtliche Erregung und Coitus. Ssowremennaja medicina. 1862. p. 916 u. 1863. p. 168 (russisch).
Weise, Priapismus de causa arthritica. Med. Zeit. in Preuss. 1853. p. 130.
Westphal, 1) Arch. f. Psych. u. Nervenk. 1869. p. 273.
2) Die conträre Sexualempfindung. Arch. f. Psych. 1870. II. p. 73.
3) Zur conträren Sexualempfindung. Arch. f. Psych. 1873. Bd. III. p. 225.
Wojenno-medicinski Journal. 1849. No. IV. p. 4. Ueber pathologische Perversität des Geschlechtssinnes (russisch).
Zacchias, Paulus, Quæstiones medico-legales. Lib. IV. T. II. Lugduni. 1726.

INDEX.

Abscesses, — sign of sodomitic rape, 195
acquired pederasty, 5, 91, 204
active pederast, 23, 56, 93, 130
active pederasty, signs of, 198
acute mania, 16
Albert, Dr., 23
alcoholic automatism, 86
alcoholism, 86
Alps, — effect of climate, 47
ambitious mania, 83, 123
anal folds, radial, 172, 173, 174
anamnesis, 205
Anjel, 69
Anthony, Saint, 88
Antinous, 139
anus infundibuliformis, 167, 168, 169
Armenian mountains, — cradle of pederasty, 47
Asiaticus, 139
athletic exercises, 20
atony of sphincter ani muscle, 174—177
" aura ", the — in epilepsy, 70

Barracks, — favourable to pederasty, 96
Bartle, Carl (of Augsburg), — case of, 32

Basse des Remparts, Rue (Paris), — case in, 56
bathing establishments at St. Petersburg, — favour pederasty, 145
Bertrand, Sergeant, — case of, 37
bestiality, 1
birds, sexual act with, 112
black-mailers, 56, 145, 149, 154
Blumröder, 37
boarding-schools, 91
Brierré de Boismont, 67
Brouardel, 36, 161
Buffon, 88

Cæsar, Julius, 137
Calvis, Marshal de, 42, 47
Canler, Mémoires de, 149
cantharides, 89, 126
Caracalla, 117
Casper, 156, 157, 161, 173, 192, 196
chancre on penis, 124
Charcot and Magnan, 26, 50
Charenton, Lunatic Asylum at, 102
China, pederasty in, 142
Church dignitary, — case of necrophilism, 66

Clysophus, 84
coitus per anum, 135
coitus per anum et os, 147
coitus per feces [?], 134, 135
College-Registrar, — case of, 63
complex forms of genesic perversion, 129
congenital contra-sexual feeling, 8
congenital cynedes, 83
congenital pederasts, 102
congenital pederasts,
— feminine appearance and attire, 13
— predisposition to feminine work and manners, 17
congenital pederasty, 2, 8, 44, 50, 68, 204
conoidal sinking between nates,
— sign of sodomy, 191
constrictor cunni muscle, 170
contra-sexual feeling, 8
Cordillera, — effect of climate, 47
Cornil, Dr., 159
corpses, defiling human, 38, 39
(see also *necrophilism*)
Cours, Abbé de, 88
cretinism, 44
criminals, — psychopaths and neuropaths *not*, 158
"crista", 190
Crothers, F. D., 86
cruelty in senile dementia, 212, 214
Cullerrier, 168
Cupid, statue of, — at Delphi, 85
Curio, — on Cæsar, 138
cynedes, 17, 53, 84

(see also *pederasts, pederasty, sodomy*)
cynedes, congenital, 83
cynedes, detection and diagnosis of, 160, 161
cynicism of language, — in old men, 108

Decrepitude, 5
defecation, 101, 110, 111
(see also *stercoraires, les*)
degeneracy, 24, 28, 32, 44
degeneracy, psychical, 24, 37, 44, 47, 53, 71, 72, 73, 80
delirium, 88, 113
Delphi, statue of Cupid at, 85
Démeaux, 45
Demme, 32, 34
detection and diagnosis of cynedes, 160, 161
Diez, C. A., 30
dipsomania, 52, 59, 158
disease and vice, 217, 219
disproportion in age of parents, 45
Doryphorus, — Nero's freedman, 140
dotage, dementia of, 106
(see also *senile dementia*)
droshky-drivers at St. Petersburg, pederasty among, 145, 147
drunken habits of parents, 45

East, pederasty in the, 103
ecchymosis, — sign of sodomitic rape, 194
Elisa Edwards, case of, 16
"enfesser", 133
engineer of Trieste, case of, 61
epilepsy, 4, 71
epilepsy, hereditary, 73

epilepsy, psychical, 5, 70
epilepsy, — the "aura", 70
epileptic crisis, 72
epileptic fits, 24
epileptic mania, 77
epileptic pederasty, 5, 71
erections, nocturnal, 10
erethism, 11, 20, 21, 25, 125, 131
erethism, exaggerated, 50, 73, 74, 80, 210
erethism, maniacal, 65
erethism, — excited by
— contact with fur, 26
— nails in women's shoes, 25
— night-cap, 24
— white apron, 25
— white objects, 31
Erlicki, Dr., 76
erotic delirium, 84
erotic fever, 84
erotomania, 5, 42, 79, 83
erysipelas, 64
Espallac, Dr., 197
Evrard, Dr., 159
exhibitionists, 57, 109
exposure, indecent, 57

Fashion in sexual depravity, effect of, 100, 101
fellators, 132, 203
feminine appearance and attire of congenital pederasts, 13
feminine manners and work, predisposition to, — among congenital pederasts, 17
Ferris, N. C. and J. C. Shaw, 50
fishes, little, — Tiberius', 138
fistulas, — sign of sodomitic rape, 179, 195

flagellation, 21, 22, 52, 53, 54, 55, 102, 103, 115
Flemming, 45
forensic medicine, 155
fur, contact with, — an incitement to erethism, 26

Genesic instinct, 1, 4
"genesic mania", 89
genesic perversion, complex forms of, 129
genitals, — irregularly developed, 28
genito-spinal centre, irritation of, 126
Giraldès and Horteloup, 197
Gock, von, 65
gonorrhœa of rectum, — sign of sodomy, 188

Hadrian, 137
hallucinations, 28, 74, 88
Heliogabalus, 139
hemorrhoidal nodes, — sign of sodomy, 179
Himalayas, — effect of climate, 47
hyperæsthesia, 209
hypochondria, 28
hysteria, 28

Idiocy, 47
incontinentia alvi, — sign of sodomy, 180
indecent exposure, 57
India, pederasty in, 142
Inebriate Automatism; F. D. Crothers, 86
infundibuliform widening of anus (anus infundibuliformis), 161, 167, 168, 169, 172
infundibulum vulvæ, 170

inspissations, — sign of sodomy, 180

Jacob, S., 116
Japan, pederasty in, 142
Jésus, les petits, 149
Juvenal, 141, 190

Kowalewski, Dr. P. J., 77
Krafft-Ebing, 36, 61
Krauss, Dr. A., 29, 32, 34, 148

Lacerations, — sign of sodomitic rape, 195, 196
Lasègue, 36, 57, 109
Laval, Gilles de, — case of, 116
Leger, 88
Legrand du Saulle, 79
levator ani muscle, 166, 168
locomotor ataxy, 127, 128
Lombroso, Prof., 31, 32, 37
Lorry, 84
love philters, 89
"lumen" of blood vessels, narrowing of, 107

"**Madmen** at liberty", 142
Magnan (and Charcot), 26, 50
malformations, 28
mania, 88
mania of persecution, 74, 83
mania, acute, 16
mania, ambitious, 83, 123
mania, epileptic, 77
mania, erotic, 5, 42, 74, 83
mania, genesic, 89
mania, periodical, 71
manual examination, 175
"marisene", 190
Marius, Caius, 137

Marquisi, Dr., 197
Martial, 141
Martineau, 169
masturbation, 12, 19, 22, 23, 26, 35, 38, 43, 55, 73, 76, 80, 97
(see also *onanism*)
masturbation, mutual, 12
Maudsley, 219
megalomania, 113
megalomania, religious, 74
melancholy, 58
Menesclou, — case of, 36, 159
Michelet, 221
Mierzejewski, Dr. W., 14, 44, 63, 116, 150, 196
Moraud, 189
morbid depravation, 219, 221
Moreau de Tours, P., 87, 89
Motet, 36

Nails in women's shoes, — an incitement to erethism, 25
necrophilism, 37, 38, 39, 53, 66, 67, 68
necrophilism, case of, — in Church dignitary, 66
Negris, Dr., 123
Nero, 140
nervous centres, 8, 9
nervous maladies, 44
nervous system, 9, 28
neuropathic constitution 3, 46, 68, 72, 87
neuropaths *not* criminals, 158
night-cap, — an incitement to erethism, 24
nocturnal erections, 10
nocturnal pollutions, 10, 23, 26
Nouvelle Justine, La; Marquis de Sade, 102

Numa Numantius, 14

Occasional pederasts, 97
œdema of prepuce, — sign of sodomitic rape, 196
old men, morbid taste for, 24
onanism, 12, 21, 131
(see also *masturbation*)
onanism, pictures and statues incite to, 84
onanistic excitation, 12
orificium ani, 23, 160, 162, 178, 179

Paralysis of the insane, 118
paralysis, progressive, 206
paralytic dementia, 118
paralytic idiocy, 6, 119
paralytic pederasty, 6
paresis of vasomotor nerves, 121
Paris, — case of Rue Basse des Remparts, 56
Paris, — les petits Jésus, 149
passive pederasts, 56, 93, 104, 130
passive habitual pederasts, 163
pédérastes, Sous-brigade des, 146
pédérastie des ramollis, 118
pederastomania, 83
pederasts, active, 23, 56, 93, 130
pederasts, congenital, 102
pederasts, congenital, — feminine appearance and attire, 13
— predilection for feminine work and manners, 17
pederasts, occasional, 97
pederasts, passive, 12, 56, 93, 104, 130

pederasts, passive habitual, 163
pederasts, prostitute, 94, 95
pedernsts, prostitute, — at St. Petersburg, 145
pederasts, senile, 131
pederasty, 1, 19, 23, 47, 48, 49, 53, 65, 92, 102, 120
(see also *sodomy*)
pederasty, — favourable conditions in
— barracks, 96
— prisons, 96
— sailing-ships on long voyages, 96
— schools, 92, 93, 94, 96
pederasty in
— China, 142
— India, 142
— Japan, 142
— Persia, 142
— Russia, 145, 146
— the East, 103
pederasty, acquired, 91, 204
pederasty, congenital, 2, 8, 44, 50, 68, 204
pederasty, epileptic, 5, 71
pederasty, periodical, 4, 52, 206
pederasty, senile, 6, 104, 106, 116
pederasty, senile, — Schopenhauer on, 114
pederasty, signs of, 160
pederasty, signs of active, 198
periodical mania, 71
periodical pederasty, 4, 52, 206
periodical perversion of genesic sense, 4
periodical psychosis, 4
persecution, mania of, 74, 83, 113
Persia, pederasty in, 142
petits Jésus, les, 149

Petronius, 141
phagedenic ulcus molle recti,
— sign of sodomy, 185
philters, 89
pictures incite to onanism, 84
"pisciculi", Tiberius', 138
Platonic love, 12, 81
pocula amatoria, 89
"podice laevi", 172
pollutions, 20, 23
pollutions, nocturnal, 10, 26
"pompeurs de dard", 132
priapism, 125, 126, 127
priapism, acute, 90
prisons, — favourable to pederasty, 96
procurers, 56
prolapsus recti, — sign of sodomy, 180
prostitute pederasts, 95
prostitute pederasts at St. Petersburg, 145
prostitutes, 53, 54, 119
prurigo ani, 108
psychopathic constitution, 3, 65, 69
psychopathic tendency, 31
psychopaths, 14, 85
psychopaths *not* criminals, 158

Radial anal folds, 172, 173, 174
Raggi, Dr., 86
ramollis, pédérastie des, 118
ramollissement (softening of the brain), 37
Rayes, Gilles de Laval, Sieur de, 116
Registrar, College, — case of, 63
remorse, 58
"renifleurs, les", 111

Riche de la Popelinière (Le), 103
Roman Emperors, 136
Rousseau, J. J., 21
rupture of frenulum, — sign of sodomitic rape, 196
Rurot, 45
Russia, pederasty in, 145

Sade, Marquis de, 102
sailing-vessels on long voyages, — favourable to pederasty, 96
St. Petersburg, pederasty at, 145
satyriasis, 5, 87
schools, pederasty in, 92, 93, 94, 96
Schopenhauer, 114
senile cynicism of language, 108
senile dementia, 101, 102, 106, 116, 206, 210, 211, 214, 218
senile dementia of Tiberius, 138
senile dementia, cruelty in, 212, 214
senile pederasty, 6, 116, 131
Seweke, Dr., 162
sexual aberrations under Roman Emperors, 136
shame in presence of men, 9
Shaw, J. C. and N. C. Ferris, 50
"sniffers", the, — les renifleurs, 111
sodomitic rape, 193
sodomitic rape, signs of,
— abscesses, 195
— ecchymosis, 194
— fistulas, 179, 195

— lacerations of scrotum and perineum, 196
— lacerations round anus, 195
— œdema of prepuce, 196
— rupture of frenulum, 196
sodomy, 1, 12, 35, 48, 49, 56 (see also *pederasts, pederasty*)
sodomy, signs of,
— conoidal sinking between nates, 191
— gonorrhœa of rectum, 188
— hemorrhoidal nodes, 179
— incontinentia alvi, 180
— inspissations, 180
— phagedenic ulcus molle recti, 185
— prolapsus recti, 180
— syphilitic induration, 181
— verrucose excrescences, 179
softening of the brain, 118
Sous-brigade des pédérastes, 146
sphincter ani muscle, 164, 165, 166, 168
sphincter ani, atony of the, 174, 175, 176, 177
Sporus, — Nero's freedman, 140
statues incite to onanism, 84
"Stercoraires, les", 101
Suetonius, 117, 138, 141
suicide, inclination to, 28
syphilis, 45, 46
syphilitic induration, — sign of sodomy, 181
syphilitic infection, 136

Tabes dorsalis, 127, 128
Tardieu, 38, 42, 111, 132, 156, 173, 196, 197
Taxil, Leo, 66, 101
Taylor, 16
theft, inclination to, 28
theft, inclination to, — of women's wash-linen, 29
Tiberius, 117, 138
Trajan, 140
Trieste, engineer of, — case of, 61

Ulrichs, K. H., 14
urethritis, 126, 136
uro-genital system, affections of, 126

Vasomotor nerves, paresis of, 121
vasomotor system, 119
venal cynedes, 124
venal passive pederasts, 94
Venus of Milo, 85
verrucose excrescences,
— sign of sodomy, 179
vice and disease, 217, 219
Vitellius, 138
vulvæ infundibulum, 170

Wash-linen, women's — inclination to steal, 29
white apron incites to erethism, 25
white objects incite to erethism, 31
Witz, Dr. M., 54
wounding young girls, 32

Zacchias, 172, 189

LIST OF Mr. CHARLES CARRINGTON'S Recent Publications.

Important new work in English from the original of
Dr. B. TARNOWSKY
(Of the Imperial Academy of Medecine, St. Petersburg.)

THE SEXUAL INSTINCT
and its Morbid Manifestations
FROM THE DOUBLE STANDPOINT
of Medical Jurisprudence and Psychiatry
DONE INTO ENGLISH NOW FOR THE FIRST TIME

BY

Messrs. W. C. COSTELLO PH.D.

AND

ALFRED ALLINSON, M.A. (Oxon)

with a short Preface written expressly by the eminent Author for this edition.

Printed on stout vellum paper and strongly bound in English cloth together with an **Analytical Index** and a **Bibliography**.

Price: SEVEN SHILLINGS and SIX PENCE, net.
(DELIVERED FREE BY POST)

N.B.—This work is intended only for Medical Men, Lawyers, and Students of Mental Pathology and should **not** be placed into the hands of the general public.

Dr. CABANÈS'
CABINET OF HISTORY.

The Medical Standard (*Chicago*):—"To the casual observer this book might seem to be worthy a place only in the Index Expurgatorius, but to the student of history and to the book-lover it is a rare prize. The bearing of things medical on the history of nations is, as a rule, carefully concealed from public view; but Dr. Cabanès has opened the secret cabinet and disclosed its interesting contents. The book is prefaced by a letter from M. Sardou, in which he courteously declines to write a preface.

"History, says Macaulay, is made up of the bad actions of extraordinary men.... Nine-tenths of the calamities which have befallen the human race had no other origin than the union of high intelligence with low desires. The Secret Cabinet of History is made up of descriptions of the extraordinary actions of great men and women.

"From these little glimpses behind the scenes, the reader will gain many ideas of the inherent divinity of kings which may be at wide variance with his preconceived opinions. The book-lover will be delighted with the appearance, paper and artistic make-up of the book. We shall await with interest the second series, which is soon to be published."

NEW AND CONSIDERABLY ENLARGED ENGLISH EDITION.

Two Stout 8vo Volumes of about 450 Pages each

(Size, 6 by 9½ in.), printed on STOUT VELLUM paper specially manufactured for this Edition by VAN GELDER, tastefully bound in strong black cloth, gilt top, untrimmed edges. This Edition is accompanied by a Set of 24 Illustrations on fine PAPIER COUCHÉ executed by

DRAEGER (of Paris), after the Original Aquarelles of
AMEDEE VIGNOLA.

Price of the Two Volumes and 24 Illustrations £2. 10s. (nett).

Untrodden Fields

of Anthropology.

OBSERVATIONS ON

THE ESOTERIC MANNERS AND CUSTOMS OF SEMI-CIVILISED PEOPLES.

Being a Record by a French Army Surgeon of Thirty Years' Experience

in ASIA, AFRICA, AMERICA, and OCEANIA.

WITH SPECIAL REFERENCES TO THE WORKS OF

SIR RICHARD F. BURTON, LOMBROSO,   MANTEGAZZA,
HAVELOCK ELLIS,   DR. PLOSS,   KRAFFT-EBING,
ARCHDN. GRAY (of Hong-Kong), DR. SCHLEGEL, EDWARD B. TYLOR,
            HERBERT SPENCER, AND OTHERS.

Further Descriptive Circulars may be had, price 6d. (for Postage).

☞ The Second volume contains a full **Analytical Index** and a **Complete Bibliography,** and the Two volumes contain TOGETHER about **Five Hundred** pages **extra** of entirely **new** matter **not contained** in the FIRST EDITION.

A Phase of Historical Science still in its Infancy.

CURIOUS BYPATHS OF HISTORY.

Studies of LOUIS XIV; RICHELIEU; MLLE. DE LA VALLIÈRE; MME. DE POMPADOUR; SOPHIE ARNOULD'S SICKNESSES; THE TRUE CHARLOTTE CORDAY; A SAVAGE "HOUND"; IN THE HANDS OF THE "CHARCUTIERS"; NAPOLEON'S SUPERSTITIONS; A PHYSIOLOGICAL PROBLEM (MME. RÉCAMIER AND QUEEN ELIZABETH OF ENGLAND), etc., followed by a fascinating study of

Flagellation in France

FROM A

MEDICAL and HISTORICAL STANDPOINT.

With special FOREWORD by the Editor dealing with the Reviewers of a previous work, and sundry other cognate matters good to be known; particularly concerning the high-handed proceedings of BRITISH PHILISTINISM.

A fine Copper-Plate Frontispiece after a Design by DANIEL VIERGE (engraved by F. MASSÉ)

The whole in two volumes on specially made, stout, white (VAN GELDER) vellum paper.

Price £1. 10s.

N.B.—With this book is given a fine Plate entitled: CONJUGAL CORRECTION reproduced in AQUATINTE by the MAISON GOUPIL (of Paris) after the famous OIL PAINTING of CORREGGIO.

This book appeals to all classes. The Scholar will find a mine of information mostly new; the Bibliophile will prize it for its "get-up"; and the "general reader" will be struck with its quaint old-world charm and real curiousness.

Recent Opinions of the English Press on

CURIOUS BYPATHS OF HISTORY

By Dr. CABANÈS
(of the Medical Faculty of Paris)

Reynolds's Newspaper.

"This is a companion volume to the work by the same author recently noticed in this column, 'Secret Cabinet of History'. It is even more interesting than that remarkable book. It might be aptly entitled a 'Chronicle of Historical Scandals', though several of the sketches are aimed worthily to show that scandal was wrongly associated with certain names, such as those of Robespierre and of Charlotte Corday. Others, however, do not come so well off—George Sand and her lovers Alfred de Musset and Dr. Pagello, Mlle. de la Vallière, Madame de Maintenon and Madame de Montespan, mistresses to Louis XIV., Madame de Pompadour, the mistress of Louis XV. Interesting contributions are 'The Superstitions of Napoleon I.,' and 'The Peregrinations of the Body of Richelieu'. Altogether we have here one of the most singular, quaint, out-of-the-way collections of odd facts and anecdotes, most of which have never appeared before, about persons who have loomed largely on the world's stage, ever put together. And the importance of this kind of knowledge is the revelation of the extent to which the domestic affairs of the great affect the policies of nations. The translation and curious notes are done into admirable English."

Pall Mall Gazette.

"About six months ago it fell to my lot to discuss in these columns a volume entitled 'The Secret Cabinet of History' (Paris: CARRINGTON), and I find myself on the horns of a similar dilemma. The volume is undoubtedly interesting, and deserves, on its merits, to be treated with every consideration; at the same time, it is impossible to discuss it fully in a journal which is likely to fall into the hands of 'the young person'. In his Foreword the editor, quoting an Hibernian friend, complains with great bitterness of 'the shout of ominous silence' with which the former work was received by the English press, and of the narrow, puritanic spirit in which most of the journals which did notice it treated it. It seems to me that when a publisher insists on publishing works which, whether rightly or wrongly, give offence to our ideas of morality, he ought to grin and bear it without vituperation if the reviewers give it a reception which they think reflects the views of the majority of their readers.... Now although the pathology of history which forms the subject-matter of Dr. Cabanès' essays is a very fascinating study, and, as being calculated to throw a new light on many obscure problems, more especially in modern history, one that the scientific historian is bound to take into account it is essentially strong meat, suitable for the student only and not fitted for the consumption of 'babes and sucklings'. For example, it is undoubtedly instructive to learn that Louis XIV. was suffering from raging toothache at the time of the revocation of the Edict of Nantes; but the doctor in history, frequently and of necessity, trenches on subjects which are not publicly discussed in this country. At the same time, it is impossible not to recognize the value of Dr. Cabanès' researches, or to appreciate the accuracy, industry, and sincerity with which they have been conducted. The papers on Charlotte Corday and on the superstitions of Napoleon throw the light of medical science on pathological characteristics which it behoves the historian to take into account, and, moreover, to most of these essays the most prurient-minded critic can take no exception.... Nor is there any obvious reason for raking up, in another essay, the scandal of George Sand's amours. Apart from these blemishes, this volume is suggestive and interesting more especially as a contribution to a phase of a historical science which is still in its infancy; but it is emphatically a work for the library and not for the boudoir or the school-room."

This work is strictly limited to 500 copies, all press-numbered; the type has been distributed and the book will not be reprinted.

Tastefully bound in black cloth, gilt top, untrimmed edges, gold lettering on back and sides.

The Mysteries of the Past laid open.

Medico-Historical Glimpses into the Private Lives of Kings, Queens, and other high Persons.

The Secret Cabinet of History.

PEEPED INTO BY A FRENCH DOCTOR.
(DR. CABANÈS).

Englished by W. C. COSTELLO, and preceded by a letter from the pen of M. VICTORIEN SARDOU.

(DE L'ACADÉMIE FRANÇAISE).

Medicine (HAROLD N. MOYER, M.D., Editor, DETROIT, Mich., U.S.A.):—"Pathology in history is always of interest, especially since by affecting the leaders or rulers of a nation it tends to affect the nation a as whole. This volume contains thirteen separate essays on subjects involved in historical pathology. The first one relates to a youthful indiscretion of Louis XIV. whereby, as Dr. Cabanès proves, Louis XIV. very early contracted gonorrhea. While it was recognized by his courtly physicians, they concealed the origin of the disorder from interested motives, although describing its clinical features very clearly. The second discusses the fistula of the same king, and mentions all the surgical procedures adopted as well as the fees paid to surgeons. The disease became fashionable among the courtiers, and one lady of honor was disconsolate because her physician was unable to discover that she had any trace of the royal disorder. The phymosis from which Louis XVI. suffered played a large part in determining certain features of the French Revolution. For a while it rendered Louis XVI. impotent and this fomented intrigues by Louis XVIII., Charles X., and the Duke of Orleans, which seriously damaged the moral character of Marie-Antoinette. These intrigues undoubtedly aided the general revolutionary movement, since all three of the intriguants hoped to profit by this movement. Another item of special interest to physicians is the proof given that Marat suffered from both mental, dermic, and other somatic symptoms of diabetes, which doubtless underlay his suspicional and pitiless tendencies. The change recorded in Marat's attitude during the Revolution is readily explained by the tone given his thoughts through the irritability produced by this disease. The work of which this is the first volume merits purchase and perusal by any physician of literary and historical tastes. The second volume will prove equally interesting, to judge from the table of subjects given in the present volume."

The Maryland Medical Journal:—"The inside history of great persons is always a matter of interest, and in these days, when the history of medicine is attracting so much attention, it is extremely instructive to note how the illness or indiscretion of crowned heads has affected so materially the history of the world. The therapeutics of the persons treated in this book are extremely crude, but the surgical procedures are in places very praiseworthy.

"The gonorrhea of Louis XIV. is well described by his physicians with great accuracy, but the treatment was not very effective. Hence during the course of the disease the physicians and courtiers took great pains to conceal the nature of his trouble from the nation. This same king's fistula caused great consternation among his attendants, and the operative procedures, after all other means had been exhausted, were indeed exceedingly creditable. The remuneration of the surgeons was rather startling.

"The remaining chapters in this very interesting work are as follows: The Maladies of Louis XV.; The Semi-Impotence of Louis XVI.; The First Pregnancy of Marie-Antoinette; Louis XVI. in Private Life; One of the Judges of Marie-Antoinette, the Surgeon Souberbielle; What was Marat's Disease; Talleyrand and the Doctors; The Accouchement of the Empress Marie-Louise; The Ancestors of Marshal Mac-Mahon, and Gambetta's Eye.

"The phimosis and impotency of Louis XVI. and the intrigues of Louis XVIII., Charles X. and the Duke of Orleans undoubtedly had their effect on the French Revolution. In all these chapters it is shown that royalty may be honoured and revered by the people, but few men are great in the eyes of their physician.

"This work is well translated and clearly printed, and will be followed by others of similar character, the prospectus of which may be obtained from the publisher. While such works are, as a rule, for physicians only, this work is written in a very nice style, and could in no way affect the delicate sense of a sensible person."

The Pall Mall Gazette *(June 5th, 1897)*. DOCTORS IN LITERATURE:—"The medicine man has of recent years been infected by the *cacoëthes scribendi*. Heretofore, he was apparently inoculated with some lymph which rendered his verbiage intelligible to the layman and confined his writings to the medical journals. Since Dr. Conan Doyle escaped, a good many of his colleagues seem to have broken away invaccinated. Hence they make incursions into fiction, and generally with some success, for the doctor of to-days is, as a rule, a man of a liberal education. We now have the doctor as an historian in M. Cabanès's 'Secret Cabinet of History' (Paris: CARRINGTON), and very interesting he is. The side-lights the man of medicine throws on certain obscure points of French history are of real, if somewhat scurrilous, value. After all, the personal factor—more especially in the mediæval history—cannot be altogether eliminated. Professor Ihne was fond of saying that if Charles V. at one critical period had not had such a bad attack of gout the history of the Reformation in Germany would probably have been very different. Very amusing is the account of the maladies of Louis XV., and, as the translator points out, the Well-Beloved must have had a wonderful constitution to have withstood all the remedies—the tobacco pills, the two glasses of manna, and the 'spirits of crayfish'—a dozen physicians poured into him from time to time. The horrors, too, of the deathbed scene in the Œil de Bœuf as painted by a doctor, surpass even Carlyle's lurid picture. Indeed all the essays in this little volume are of interest, full of quaint learning and deep research. The fate of the brain of Talleyrand, 'that brain', as Victor Hugo cried, 'which had inspired so many men, built up so many schemes, led two revolutions, deceived twenty monarchs, and had contained the world', and at the last was thrown into the drain of the Rue Richepanse, and the quest of the lost eye of Gambetta, are among the striking curiosities of history. At the same time it must be added that the volume is not written *pueris virginibusque*. Although the author is anxious not to outrage what the translator sweetly calls the 'pudicities', his resolve is at times swept away by the interest of his studies. Moreover, the original was written in French. Lest it fall into the hands of a maiden uncle the 'Secret Cabinet of History' should be kept under lock and key along with Balzac's 'Contes Drolatiques'; to which, by the way, it forms a sort of scientific commentary."

Reynolds's Newspaper *(May 30th, 1897)*:—"This is a very curious book, as the titles of some of the chapters will indicate. Thus: 'A Youthful Indiscretion of Louis XIV.', 'The Fistula of a Great King', 'The First Pregnancy of Marie-Antoinette', 'What was Marat's Disease'? 'Gambetta's Eye', 'The Semi-Impotency of Louis XVI.', 'The Accouchement of the Empress Marie-Louise' The author treats of the excesses and sensualities of the Bourbon monarchs of the pre-Revolution days—Louis XIV., diseased from his corrupt youth; the filth and immoralities of Louis XV.; the cruelty, sordidness, and stolidity of Louis XVI.

"'The Secret Cabinet of History' is a book full of the most singular and out-of-the-way knowledge about distinguished Frenchmen and women for a period of about a hundred years. The information given by this French physician is generally not accessible to the ordinary historian; but it is valuable as showing how much of the changes in the government of peoples depends upon the vices, frailties, and diseases of rulers—dispositions to which they seem more largely to succumb than do ordinary citizens."

Jamaica Post *(Thursday, July 6th, 1897)*:—"So thorough has been its success in the original (French), that an edition in English has now been brought out by the enterprising publisher. And that equal success will attend it wherever it penetrates throughout the British Empire and the United States, is a foregone conclusion. So far as our purview goes, there is nothing in literature with which it can be compared. Something (although not very much) in the nature of G. W. M. Reynolds's notorious 'Mysteries of the Court of London', there might be a parallel between the two works, save for the fact that whereas the 'Mysteries' is almost pure fiction, Dr. Cabanès's book is obviously pure, or rather impure (*very impure!*) history. The present volume is but the first of a series and the published contents of the forthcoming second volume give promise of equally entertaining matter."

The following works are in the press:

History of the Plague of Lust in Classical Antiquity, including:—Detailed Investigations into the **Cult of Venus,** and **Phallic Worship,** Brothels and Public Women, the Νοῦσος Θήλεια (*Feminine Disease*) of the **Scythians, Pæderastia,** and other **Sexual Perversions amongst the Ancients,**—As Contributions towards the **Exact Interpretation of their Writings** by DR. JULIUS ROSENBAUM. Translated now for the *first time* from the Sixth (*Unabridged*) German Edition with all the Latin and Greek passages **done literally into English** by ALFRED ALLINSON, M.A. (Oxon). The whole with full **Bibliography** and **Index** in *One stout volume* of some **600 pages** (size 6 × 9 inches), strongly bound in English cloth.

The Dangers of Debauchery, with especial reference to the Intellectual and Physical Faculties, its Influence on the Health and Human Life, translated from DR. VIREY, with Annotations.

Medical Studies of the Latin Poets, translated from the French of Dr. P. MENIÈRE, together with considerable additions and textual illustrations.

La Jeunesse rendue aux Vieillards (*Liber redintegratæ Aetatis in potentia Libidinis*):—traduction faite sur les **MSS. arabes** (1063 de l'Hégire) dans la Bibliothèque Nationale à Paris. Suivie d'un Supplément traitant de la **Nature** et **Efficacité** des **Aphrodisiaques.**

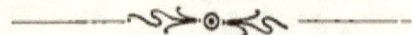

CHARLES CARRINGTON
PUBLISHER OF MEDICAL, HISTORICAL, AND FOLKLORE WORKS,
13, FAUBOURG MONTMARTRE, PARIS.

Mr CHARLES CARRINGTON begs to call the attention of the English reading public to the following virile and realistic story (about 260 pages 8vo). This book is beautifully printed on good vellum paper and its charming "get up" is sure to please all classes of readers.   Price **10s.**

The Chastisement
OF
Mansour

BY
HECTOR FRANCE

Done into English by ALFRED ALLINSON.

In an eloquent "Postcript" the Author says:—"A weekly Review bearing the curious title of 'The Bat', a name it presumably deserved, declared the book

"**The Apotheosis of Rape with Violence**",

and called for the intervention of the Public Prosecutor!.... How well we know them, the folk who veil their faces in consternation before a work of Art, because it displays undraped the glories of the human form divine! How well we know them, the folk, who in public profess themselves shocked to hear the word and all the while in secret delight to do the thing, like old maids over their tea, blushing at the very mention of such a dreadful word as *chemise*, but regaling their prurient imaginations with the pictures of passion certain chapters of the Bible present, and licking their lips over the *Song of Solomon!* But England has no monopoly of such-like oddities. 'Their name is Legion in "godfearing Germany" as it is in "immoral France"'. The tribe of *Tartufes* is just as much alive and to the fore now as it was in Molière's time, and his: **Cover up that bosom that J dare not contemplate** is quite in the modern taste."

www.ingramcontent.com/pod-product-compliance
Lightning Source LLC
Chambersburg PA
CBHW031936230426
43672CB00010B/1941